Collector's Guide To

Barbie DOLL

Paper Dolls

Identification & Values

Lorraine Mieszala

COLLECTOR BOOKS

A Division of Schroeder Publishing Co., Inc.

Searching For A Publisher?

We are always looking for knowledgeable people considered experts within their fields. If you feel that there is a real need for a book on your collectible subject and have a large comprehensive collection, contact Collector Books.

On The Cover

Barbie paper doll and Solo in the Spotlight outfit from
1962 *Barbie and Ken Cut-Outs,* $135.00

Western Barbie Paper Doll from 1982, $20.00

SuperStar Barbie Paper Doll from 1989, $20.00

Barbie Doll Cut-Outs from 1962, $135.00

Barbie and Ken Paper Dolls from 1971, $50.00

Cover Design: Beth Summers
Book Layout Design: Karen Geary
Photography: Jean Tracz

Printed in the U.S.A. by Image Graphics, Paducah, KY

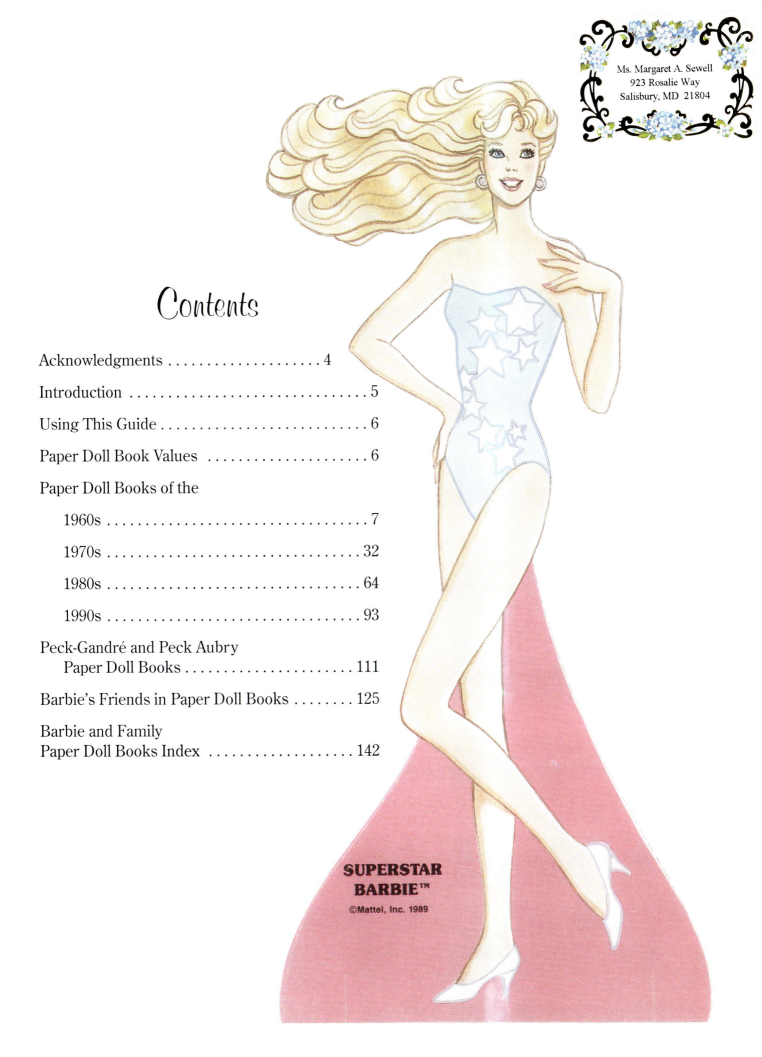

Contents

SUPERSTAR
BARBIE™

©Mattel, Inc. 1989

Acknowledgments

Thank you to my dear husband Ed, for all the support and encouragement — and a special thank-you for locating and buying the Barbie paper doll books for my collection.

Thank you to my dear daughter Karen, for always being there when needed and for all her help in every way. To my friend and photographer Jean, for her inventiveness, creativity, and patience. To Shirley for the loan of the Barbie dolls, and being a special friend. To my Barbie doll Buddies — Sandra, Mary, Elba, Elaine, and Eddie.

Introduction

The Barbie doll is an incredible phenomenon today. Widespread interest exists in every aspect of Barbie doll collecting, including her paper doll books. These paper doll books are a visual history of the world's most popular fashion doll, Barbie, the other dolls in her family, and their outfits and accessories. Some paper dolls are precise replicas of the vinyl dolls of a particular era. Other books feature paper dolls with different facial expressions and realistic body positions different from the vinyl dolls. The paper doll clothing chronicles the changes in Barbie doll's wardrobe and has the same focus on exceptional quality and fashion detail as the original clothes. These paper doll books are truly a study of popular American fashion as interpreted by Barbie doll's styles over three decades.

Collector's Guide To Barbie® Doll Paper Dolls is a comprehensive reference to these Barbie and family paper doll books, identifying and describing over 100 paper doll books sold between 1962, shortly after the Barbie doll was introduced, and the present. Divided into decades, with the cover, paper dolls and outfits of each book described in detail, this guide also compares the paper dolls and outfits to the vinyl dolls and their original outfits, highlights interesting and unusual information about each paper doll book, and provides the current market value of discontinued books.

The search for an old uncut paper doll book can be an exciting challenge. These books were intended to be cut and therefore many have been lost or destroyed. Those books which have been saved may be found at flea markets, in antique stores, or through Barbie doll or paper doll dealers. Collecting the cut pieces of a paper doll book can be challenging and enjoyable. The cut paper dolls can be dressed and displayed in frames or plastic bags or photo albums. Whatever the form, Barbie paper dolls can be a treasured additions to a Barbie collection and an opportunity to have Barbie memories at a reasonable price. The older paper doll books issued by Western Publishing are replicas of vintage Barbie dolls and outfits that are rare and sometimes prohibitively expensive. A quality paper doll collection can be assembled at substantially less expense and can be readily displayed or conveniently stored. The Barbie collector will undoubtedly find collecting the Barbie and family paper doll books an interesting and rewarding search.

Using This Guide

This book identifies all known mass-produced Barbie and family paper dolls and reprints. Additional Barbie paper doll books may be discovered. The paper doll books featured here were issued by Western Publishing Co., Inc. or Peck-Gandré/Peck Aubry. Golden Book and Whitman are trademarks of Western Publishing Company.

The guide is divided into chapters by decades from the 1960s through the 1990s. The breakdown of books by publisher is as follows: Western Publishing issued 15 books and two reprints in the 1960s, 26 books and 14 reprints in the 1970s, 22 books and 11 reprints in the 1980s, and 10 books and six reprints in the 1990s; Peck-Gandré/Peck Aubry issued three books in 1989, three books in 1994, one book in 1995, and two books in 1996.

A "reprint" is a re-issue of a paper doll book. The reprint may have a different format (folder versus book), more or fewer pages, a different cover, or a changed cover with a different price or no price. Reprints often have the original four digit stock number plus a dash and additional numbers. The reprint may appear in the same year as the original issue. In recent years, reprints have often had different covers with the same outfit pages.

A paper doll "folder" is a book with pockets in the inside front and back covers to store dolls and clothes. The words "punch-out" or "press-out" refer to dolls or outfits which have perforated outlines so they can be removed without scissors.

"Original" dolls or outfits refers to Mattel vinyl dolls and clothing. The paper dolls are often replicas of the original dolls identified in the title or pictured on the cover. When the title or cover of the book does not readily identify an original doll, the cover doll and paper doll were compared with those dolls available around the date of publication of the book to determine the name of the original doll.

Paper doll outfits are identified as replicas of original outfits when possible. An outfit that is "unknown" could not be identified as a replica and may be an artist's original or a replica of an original outfit which is unknown. The identification of color of the outfits is subjective and may be affected by the dyes in the printing process (pink and red or yellow and gold may be difficult to distinguish).

Paper Doll Book Values

Current values are based on mint, uncut books. Books which are complete but have been cut are valued at 50% of the mint value. Reprint paper doll books are valued the same as the original's issue. Value of the books is determined on the basis of rarity, age, and likeness to the original Barbie dolls and outfits. Minimal consideration was given to the independent artistic value of a book because that is a matter of subjective individual preference. Condition is the single more important factor in determining the value of a paper doll book which is not mint. True value is established by the market, where prices may vary between geographic locations (e.g., East Coast versus West Coast, smaller town versus metropolitan area).

Paper Doll Books of the 1960s

The Barbie and family paper doll books issued in the 1960s illustrate a remarkable collection of original dolls and outfits. Some of the paper dolls picture rare vintage dolls which are highly collectible, including the Ponytail, Bubble, Swirl Ponytail, Twist 'n Turn, Talking, and Standard. Barbie doll's family and friends like Ken, Midge, Skipper, Allan, Skooter, Francie, Christie, Casey, Stacey, and Tutti also appear in books with Barbie or on their own. The paper doll fashions have the same variety and detail as the original doll outfits of this era. The books contain a wide selection of beautiful evening gowns, sports clothes, casual wear, and western outfits. The most reproduced original outfit from this time is #962 Barbie-Q which is pictured in three books. Original outfits #915 Peach Fleecy Coat, #931 Garden Party, #937 Sorority Meeting, #940 Mood for Music, #975 Winter Holiday, #981 Busy Gal, and #986 Sheath Sensation are each illustrated in two books. The popular #900 series original outfits are pictured in #1971 *Barbie and Ken* and in #1962 *Midge, Barbie's Best Friend*. #1976 *Barbie, Christie and Stacey* is a great reference to the less-known Barbie and Stacey fashions.

Six books from this decade are exceptional and deserve special attention. #1971 *Barbie and Ken* has a colorful folder, heavy cardboard paper dolls of Ponytail Barbie and Crew-cut Ken, and an outstanding collection of #900 series original outfits with elegant evening gowns, day dresses with matching hats, and many accessories. #1976 *Barbie Costumes* illustrates the coveted original "Little Theater" costumes and includes five paper dolls of Barbie and family that duplicate the original dolls. #1957 *Skipper* pictures beautiful matching Barbie and Skipper outfits. #1985 *Skooter* illustrates additional Skipper original outfits (although sometimes in different colors). #1980 *Francie* has photographs of Francie dolls on the cover and includes a fascinating collection of original Francie outfits. #1976 *Barbie Has a New Look* pictures Barbie with her new long hair and straight bangs, after her late 1960s make-over, and has replicas of rare, glamorous #1600 series original fashions. This book is also unique because it has an 11" paper doll, which is oversized compared to the standard 9" paper doll.

YEAR	NUMBER	PRICE	TITLE	MAKER	VALUE
1962	**#1971**	**59¢**	**Barbie and Ken Cut-Outs**	**Whitman**	**$135.00**

An exceptional paper doll book and collector's item. An outstanding reference guide to many early Barbie and Ken outfits and accessories including Enchanted Evening, Busy Gal, Solo in the Spotlight, and Ken's Tuxedo. The book also features a beautiful and elaborate bridal gown similar to Barbie's first bridal gown.

Cover:

Blonde Ponytail Barbie #850 hugs a bouquet of pink carnations and smiles over her shoulder at a blonde Crew-Cut Ken #750 who holds one carnation.

Paper Dolls:

Two punch-out dolls: Ponytail Barbie #850 wears the original Barbie doll strapless one-piece black and white striped swimsuit. Crew-Cut Ken #750 is dressed in dark blue shorts with a white and blue striped T-shirt.

Outfits:

The cut-out paper doll outfits are "102 Go-Together Fashions for Play Fun" printed on two long sheets of paper folded into ten sections. The 102 fashions include the small accessories.

#1971 Barbie and Ken Cut-Outs

Ken and Barbie paper dolls from #1971 book

Outfits (continued):

Barbie

#961 Evening Splendour (different purse)
#962 Barbie-Q (short sleeves)
#963 Resort Set
#965 Nighty-Negligee Set
#975 Winter Holiday
#979 Friday Nite Date
#981 Busy Gal
#982 Solo in the Spotlight
#983 Enchanted Evening (light burgundy)
#988 Singing in the Shower (orange towel)
#989 Ballerina (pink)
#991 Registered Nurse

Ken

#770 Campus Hero (sweater without the letter "U"
     in red, aqua and blue with narrow white stripe)
#780 In Training
#781 Sleeper Set
#783 Sport Shorts (large pink & black circles on a
     dark green shirt)
#784 Terry Togs (green towel)
#786 Saturday Date (gray)
#787 Tuxedo

Ponytail Barbie doll with matching paper doll from #1971 book.

Accessories:

Hats, purses, belts, pearls, sunglasses, umbrella, hangers, shaving equipment, mirror, comb, brush, clock, flowers, yarn, keys, camera, record player, typewriter, radio, briefcase, telephone, hot water bottle, banjo, box of candy, milk, cookies, hot dog, orange, apple, cake, fried eggs, pie, cupcakes, milk, ice cream soda, ice skates, baseball bat, baseball, football, basketball, golf club.

Inside Cover:

Blonde Ponytail Barbie doll sits near the window of her pink bedroom, holding a telephone. Through the window, Ken can be seen in his blue bedroom talking on his phone.

Inside Back Cover:

A mirror reflection shows Ponytail Barbie doll applying her lipstick.

Back Cover:

Nine small pictures of Barbie and Ken in "go-together fashions" for "School, Prom, On the Job, Winter Sports, Barbecuing, Casual Wear, Around Town, At the Beach and Saturday Date."

Note:

The paper doll book opens to three sections with three pockets for paper dolls and outfits. One pocket contains the paper dolls. The other two pockets hold the paper doll outfits.

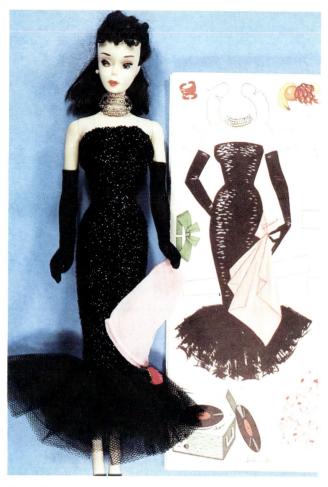

Ponytail Barbie doll dressed in #982 Solo in the Spotlight with matching paper doll outfit from #1971 book.

Ken doll dressed in #787 Tuxedo with matching outfit from #1971 book.

#1971 opened, showing the sheets of outfits to cutout. Note inside cover and inside back cover.

YEAR	NUMBER	PRICE	TITLE	MAKER	VALUE
1962	#1963	29¢	**Barbie Doll Cut-Outs**	Whitman	$135.00

A valuable book because of the pretty cover and attractive clothing from the '60s, although unlike Barbie original outfits.

Cover:

Blonde Bubble Barbie #850 walks through a city carrying her model's round yellow hat box. Barbie holds the large blue picture hat on her head.

Paper Doll:

One punch-out doll: Bubble Barbie doll #850 dressed in a white, strapless one-piece swimsuit with small pink polka dots.

#1963 *Barbie Doll Cut-Outs.*

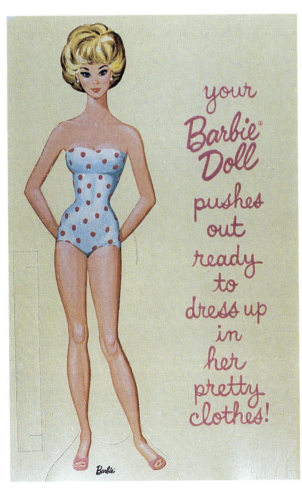

Barbie paper doll from #1963.

Outfits:

Six pages of 50 cut-out costumes that are not original Barbie doll fashions but are similar in style. The collection contains elegant evening dresses, day dresses with full and straight skirts, shorts, tops, pajamas, a swimsuit, western outfit, mink stole, small fur jacket, leopard coat, and red coat.

Accessories:

Scarves, shawls, purse, belts, necklace, bracelets, hair ribbons, fan, flowers, ice cream cone, ice cream soda, watermelon, and a goldfish bowl.

Inside Back Cover:

An open round blue hat box with a small mirror shows a reflection of Barbie putting on lipstick. The cover has a pocket to store dolls and outfits.

Back Cover:

Four small Bubble Barbie models walk in front of a city skyline view.

YEAR	NUMBER	PRICE	TITLE	MAKER	VALUE
1963	#1962	29¢	Barbie Cut-Outs	Whitman	$135.00

The Barbie paper doll is the same as the paper doll in book #1963 except Barbie is a brunette and wears a different swimsuit. Many of the Barbie and Ken doll outfits shown are different in color from the original doll outfits.

Cover:

A large close-up of a wide-eye brunette Bubble Barbie #850 who poses with her arms crossed before her. In the background three small Barbie dolls pose in different outfits.

Paper Doll:

One punch-out doll: Brunette Bubble Barbie #850 wears a one-piece, strapless aqua swimsuit trimmed in white.

Outfits:

Six pages of cut-outs with "48 Costumes and Accessories."
#850 Bubble Barbie Doll red swimsuit
#915 Peach Fleecy Coat
    (olive green with round collar and cuffs)
#931 Garden Party (solid pink with white ruffle
    and cap sleeves)
#937 Sorority Meeting
#940 Mood for Music (pink with skirt)
#962 Barbie-Q (yellow dress with red trim;
    apron in royal blue)
#969 Suburban Shopper
#981 Busy Gal (red blouse with white dots)
#986 Sheath Sensation (blue)

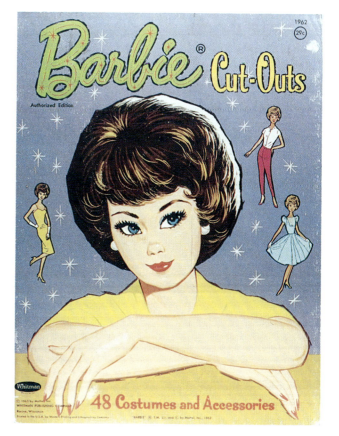

#1962 *Barbie Cut-Outs.*

Outfits (continued):

Paks (1962-1963):

Campus Belle	Silk Sheath
Gathered Skirt	Tee Shirt and Shorts
Plain Blouse	Pajama (Peter Pan collar, trim on sleeve and collar)
Scoop Neck Playsuit	Busy Morning (blue)

Accessories:

Hats, purses, scarves, necklaces, flowers.

Inside Back Cover:

Barbie doll's open closet shows her dresses hung on a pole, boxes on a shelf, and shoes on the floor. Ken doll's picture hangs on the inside door. The cover has a pocket to carry dolls and outfits.

Back Cover:

Bubble Barbie stands in front of a full-length mirror, which reflects the back of her dress, #931 Garden Party.

YEAR	NUMBER	PRICE	TITLE	MAKER	VALUE
1963	#1962	29¢	**Midge Barbie's Best Friend Cut-Outs**	Whitman	$90.00

An outstanding collection of early Barbie outfits apart from some color changes. Delightful items are added to #940 Mood for Music (a bouquet of flowers) and #953 Barbie Baby Sits (a baby in Barbie's arms).

Cover:

A large close-up of a freckled face Midge #860 puts on her white gloves. In the background a small Midge #860 clutches a yellow bouquet dressed in Belle Dress, a pak dress.

Paper Dolls:

Two punch-out dolls : Blonde Midge #860 and a redhead Midge #860 are dressed in the original doll two-piece swimsuit. Blonde Midge has a two-piece green top and blue bottom swimsuit, while redhead Midge has a pink top and red bottom swimsuit.

Outfits: Six pages of cut-out Barbie outfits.

#915 Peachy Fleecy Coat
   (different hat and purse)
#931 Garden Party (pink)
#933 Movie Date (green and white)
#934 After Five
#937 Sorority Meeting (blue)
#939 Red Flare
#940 Mood for Music (blue skirt)
#941 Tennis Anyone
#942 Icebreaker
#943 Fancy Free

#944 Masquerade (no hat)
#953 Barbie Baby Sits
#956 Busy Morning
#962 Barbie-Q (yellow dress with green trim
   and red apron)
#966 Plantation Belle (light green)
#973 Sweet Dreams (trimmed in green)
#975 Winter Holiday
#986 Sheath Sensation (olive green)
#989 Ballerina (blue tutu)

#1962 Midge Barbie's Best Friend Cut-Outs

Midge doll with matching paper doll from #1962 Midge book.

Outfits (continued):
  Paks:
  Knit sheath with fringe (red)
  Two-piece pajama

Accessories:
  Hats, purses, ice skates, ballet shoes, baby bottle, baby rattle, lock, apple, comb.

Back Cover:
  Three Midge dolls dressed in #983 Enchanted Evening, #939 Red Flare and #951 Senior Prom. The book does not include cut-out outfits of Enchanted Evening or Senior Prom.

#1962 Midge book opened showing Midge paper doll with outfits.

YEAR	NUMBER	PRICE	TITLE	MAKER	VALUE
1963	#1976	59¢	**Barbie and Ken Cut-Outs**	Whitman	$90.00

The elaborate "drive in" scene makes this a unique paper doll book. Although the fashions are not original doll outfits, many pretty clothes are pictured including a beautiful blue gown, day dresses, tuxedo, business suit, and cold weather sports wear. The book is entitled "Cut Outs," but the dolls and outfits punch out.

Cover:
A redhead Bubble Barbie #850 and a blond Crew-cut Ken #750 lean on the windshield of a red convertible.

Paper Dolls:
Two punch-out dolls: Redhead Bubble Barbie doll #850 wears a one-piece strapless pink swimsuit trimmed with navy blue. Blond Crew-Cut Ken doll #750 wears brown Bermuda shorts and a yellow sweatshirt.

Outfits:
Six pages of punch-out outfits for Barbie and Ken.

 #786 Saturday Date (olive green suit with gold tie)

#787 Tuxedo (white coat, black cummerbund and black tie)

Accessories:
Hats, purses, scarves, necklaces, and flowers.

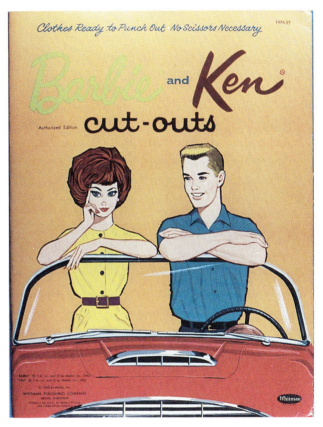

#1976 Barbie and Ken Cut-Outs

Cardboard Insert:
Punch-out a red convertible, a small sitting Barbie paper doll, Barbie doll's purse and a small Ken holding onto a steering wheel. Place the Barbie and Ken paper dolls and Barbie doll's purse with lettered tabs into slits with matching letters in the red convertible.

Inside cover:
A hamburger stand with signs, "Joe's Burgers," 10¢ hot dogs, 12¢ fries, 19¢ burgers, and 20¢ malts. The red convertible has lettered tabs to be placed into the slits of the "drive in" scene.

Back Cover:
Seven small Barbie and Ken dolls dressed in matching outfits.

YEAR	NUMBER	PRICE	TITLE	MAKER	VALUE
1963	**#1976**	**59¢**	**Barbie, Ken and Midge Paper Dolls**	**Whitman**	**$80.00**

The cover, with Barbie, Ken and Midge holding paper doll clothes is a clever idea. A delightful book because of the large size paper dolls and Ken original outfits — especially the sailor and business suits.

Cover:

Blonde Ponytail Barbie #850, blond Crew-Cut Ken #750, and Midge #860 hold paper doll clothes with tabs. Barbie holds a pink dress "Pak Belle," Ken shows his #770 "Campus Hero" pullover and redhead Midge holds a pink "Gathered Skirt" Pak.

Paper Dolls:

Three very tall punch-out dolls: Blonde Ponytail Barbie #850 wears a red and pink vertically striped swimsuit with her signature "Barbie" at the bottom. Blond Crew-Cut Ken #750 wears yellow and orange vertically striped swim trunks with a black v-neck T-shirt. Redhead freckled face Midge #860 wears a blue and green vertically striped two-piece swimsuit with her signature "Midge" at the top.

#1976 *Barbie, Ken and Midge Paper Dolls.*

Barbie and Ken paper dolls from #1976 *Barbie, Ken and Midge* book. (Note the unusual oversized paper dolls.)

Outfits:

Six pages of punch-out "52 Costumes and Accessories."

#750 Ken doll original trunks and beach jacket

#782 Casuals T-shirt

#785 Dreamboat jacket (olive green)

#786 Saturday Date (light gray)

#796 Sailor Suit, hat and bag (lighter blue bag)

#798 Ski Champion (jacket and cap)

Paks (1963):

Huntin' Shirt with rifle (white, red, and black plaid)

Sweater

Boxing shorts (light blue)

#850 Barbie doll original red swimsuit

Accessories:

Hats, purses, scarves, and belts

Inside Front Cover:

Posters hang above Barbie doll's fashion design drafting table showing outfits #934 After 5, #991 Registered Nurse, and Ken #782 Casuals. The cover has a carry pocket to store dolls and outfits.

Inside Back Cover:

Posters hang above Barbie doll's bedroom dresser showing outfits #958 Party Date, #937 Sorority Meeting, #946 Dinner At Eight, and Ken's #786 Saturday Date (gray). The cover has a carry-pocket to store dolls and outfits.

Back Cover:

"Autographed" photos of Ken, Barbie and Midge.

YEAR	NUMBER	PRICE	TITLE	MAKER	VALUE
1964	#1976	59¢	**Barbie Costume Dolls with Skipper, Ken, Midge and Allan**	Whitman	$125.00

All of the costumes are original Barbie and Ken doll Theater Costumes with a few color changes. Skipper has an Arabian Nights costume and a Poor Cinderella outfit. The paper doll book does not include Barbie doll's Arabian Nights outfit, the shield for Ken's King Arthur outfit, wolf head from Red Riding Hood (left on the back inside cover) or headpiece for Cinderella's costume from the original theater costumes. Midge and Allan can wear Barbie and Ken outfits. One of the best paper doll books and a great collector's item.

Cover:

A large blonde ponytail Barbie #850 dressed as Guinevere stands on a theater stage with a red brick background. A small Ken #750 as King Arthur, Midge #860 as Cinderella, Allan #1000 as Arabian prince and Skipper #0950 as Poor Cinderella stand behind Barbie.

Paper Dolls:

Five punch-out paper dolls: Blonde Bubble Barbie #0850 wears a strapless one-piece red swimsuit with white vertical stripes and a bow on the bodice. Brunette Crew-Cut Ken #750 wears royal blue shorts and a blue and white horizontal striped T-shirt. Redhead Midge #860 wears a one-piece dark aqua swimsuit with white ribbon trim and bows at the hips. Blonde Skipper #0950 wears her original red and white one-piece striped swimsuit. Blond Allan #1000 wears white shorts and an orange sweater with two large white horizontal stripes at the top.

Outfits:

Three pages of nine punch-out outfits and three scene pages.

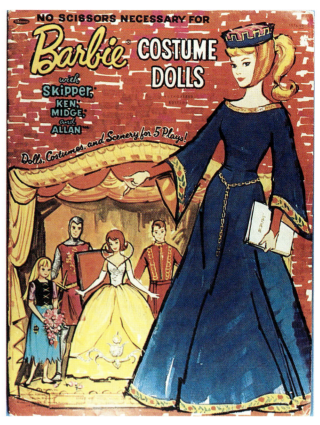

#1976 *Barbie Costume Dolls*

Barbie doll dressed in #0872 Cinderella outfit with matching paper doll outfit from #1976 *Costume* book.

Outfits (continued):

Barbie or Midge
#0873 Guinevere
#0872 Cinderella (gold skirt)
#0872 Poor Cinderella (green skirt)
#0880 Red Riding Hood

Ken or Allan
#0773 King Arthur (gold pants)
#0772 Prince (medium green cape with orange lining and orange tights)
#0774 Arabian Nights (Ken)

Skipper
Costume for King Arthur play:
#0872 Poor Cinderella (blue)

Accessories: Hats

Scenes:

Arabian Nights with genie and lamp; castle for King Arthur, Sleeping Beauty and Cinderella; forest for Little Red Riding Hood.

Inside Front Cover:

Barbie doll's theater dressing table and mirror with a message: "Barbie — rehearsal at 8AM. See you, Midge." The dressing table has a pocket to carry dolls and outfits.

Inside Back Cover:

Ken's theater dressing table, with the wolf head from Red Riding Hood and shield from King Arthur. The dressing table has a pocket to carry dolls and outfits.

Back Cover:

A red brick wall with playbills reading, Coming Soon! "Barbie as Sleeping Beauty," "Ken and Allan in King Arthur," and directions to "Make your own Theater."

Barbie paper doll outfits #0880 Red Riding Hood and #0872 Poor Cinderella and #0774 Ken Arabian Nights from #1976 *Costume* book.

#0774 Ken Arabian Nights outfit.

#1944 or #1957 *Barbie and Skipper*

YEAR	NUMBER	PRICE	TITLE	MAKER	VALUE
1964	#1957	29¢	**Barbie and Skipper**	Whitman	$80.00

All of the Barbie outfits have matching Skipper outfits except Solo in the Spotlight and Tennis Anyone. An excellent paper doll book.

Cover:

Blonde Swirl Ponytail Barbie #0850 combs #0950 Skipper's blonde hair. Barbie and Skipper are dressed in matching outfits, Barbie #0977 Silken Flame and Skipper #1902 Silk 'n Fancy.

Paper Dolls:

Two punch-out dolls: Blonde Swirl Ponytail Barbie #0850 dressed in a strapless red swimsuit trimmed with white lace. Skipper #0950 dressed in her original red and white striped swimsuit.

Outfits:

Six pages of cut-out outfits.

Barbie	**Skipper**
#0991 Registered Nurse	#0889 Candy Striper Volunteer
#0986 Sheath Sensation	#1901 Red Sensation
#0977 Silken Flame	#1902 Silk 'n Fancy
#0944 Masquerade	#1903 Masquerade Party
#0987 Orange Blossom	#1904 Flower Girl
matches Skipper's Ballet Class	#1905 Ballet Class
#0939 Red Flare	#1906 Dress Coat
#0957 Knitting Pretty (pink)	#1907 School Days
#0942 Icebreaker (red tights)	#1908 Skating Fun
#0965 Nighty Negligee (peach)	#1909 Dreamtime (peach)
#941 Tennis Anyone (gold with red dots and red trim)	
#987 Let's Dance (sleeveless)	matches Barbie's Let's Dance
#0954 Career Girl (orchid with black blouse)	matches Barbie's Career Girl
#0982 Solo in the Spotlight	

Matching Barbie and Skipper paper doll outfits from #1944 or #1957 book.

Accessories:

Hats, tennis racket, and corsages.

Inside Back Cover:

A tall flower border extends diagonally across a blue-green background. The cover has a pocket to carry dolls and outfits.

Back Cover:

Three Barbie and Skipper dolls dressed in matching outfits.

Barbie	**Skipper**
#0987 Orange Blossom	#1904 Flower Girl
#0939 Red Flare	#1906 Dress Coat
#0942 Icebreaker	#1908 Skating Fun

YEAR	NUMBER	PRICE	TITLE	MAKER	VALUE
1964	#1944	29¢	Barbie and Skipper	Whitman	$80.00

The same as #1957 Barbie and Skipper paper doll book without the inside back carry-pocket.

YEAR	NUMBER	PRICE	TITLE	MAKER	VALUE
1965	#1985	59¢	Skooter Paper Dolls	Whitman	$90.00

A valuable reference to Skipper fashions, the book contains twenty-five outfits with only nine accessories. The addition of a lei is a mystery, because none of the outfits is Hawaiian.

Cover:

Three smiling Skooter dolls #1040 standing side by side. Redhead Skooter wears #1901 Red Sensation and Blonde Skooter wears #1902 Silk 'n Fancy while the brunette Skooter wears #1904 Flower Girl.

Paper Dolls:

Three punch-out dolls: Skooter #1040 paper dolls wear the same style of two-piece swimsuit. Redhead Skooter doll's swimsuit has a blue top and blue bottom with purple and green diagonal lines. Brunette Skooter doll's swimsuit has a white top with red dots and a red bottom. Blonde Skooter doll's swimsuit has a green top and a yellow bottom with dark and light green diagonal lines.

Outfits:

Six pages of 34 punch-out outfits and accessories.

#1901 Red Sensation (jacket)
#1902 Silk 'n Fancy (white jacket)
#1903 Masquerade Party (orange and black)
#1904 Flower girl
#1905 Ballet Class (all pink, no headpiece)
#1906 Dress Coat (pink)
#1907 School Days
#1908 Skating Fun (no hat or muff)
#1909 Dreamtime (deep pink robe)
#1910 Sunny Pastels (orange with pink flowers)
#1914 Platter Party (white top, pink skirt)
#1915 Outdoor Casuals (without a collar)
#1916 Rain or Shine (gold)
#1920 Fun Time

Accessories: Purses, collar, and lei

#1985 *Skooter Paper Dolls.*

Skooter doll with paper doll from #1985 book.

Cardboard Insert:
Punch-out the handles and place into the cover to make a tote.

Inside Front Cover:
A steamer trunk covered with cushions has stickers from Rome, Pisa, and Madrid. Travel posters for Africa, India, and Tokyo hang above the trunk. The cover has a pocket to keep clothes in Skooter's trunk.

Inside Back Cover:
Photographs of Barbie, Ken, Midge, Allan and Skipper hang on a dressing screen. The cover has a pocket to store dolls behind the dressing screen.

Back Cover:
One large Skooter wears #1914 Platter Party while holding a few pink carnations. To one side sits a small, redhead Skooter wearing #1907 School Days without the sweater. To the other side, a small brunette Skooter poses in #1908 Skating Fun.

YEAR	NUMBER	PRICE	TITLE	MAKER	VALUE
1966	#1976	59¢	Barbie, Skipper and Skooter Paper Dolls	Whitman	$65.00

The idea of Barbie, Skipper and Skooter shopping together is an interesting theme. However, the clothes are simple and not original doll outfits. Each outfit page has Barbie, Skipper and Skooter signatures.

Cover:

A close-up of Bubble Barbie #0850, Skipper #0950 or #1030 and Skooter #1040 or #1120 admiring an outfit in a store window.

Paper Dolls:

Three punch-out dolls: Blonde Bubble Barbie #0850 wears a one-piece blue dotted swimsuit trimmed in purple. Blonde straight leg Skipper #0950 or bendable leg Skipper #1030 wears a two-piece swimsuit with a pink bottom and horizontally striped pink and green top. Redhead straight leg Skooter #1040 wears a two-piece swimsuit with an overall gold and orchid design on a white background.

Outfits:

Six pages of colorful punch-out outfits for Barbie, Skipper and Skooter. The outfits include basic day dresses, coats, sleepwear, and evening wear. Skipper and Skooter can share clothes. Multiple signatures of Barbie, Skipper and Skooter appear on each page.

Accessories:

Hats and purses.

Inside Front Cover:

An open wardrobe cabinet with boxes stacked on the top has dresses hung inside with price tags. The cover has a carry-pocket to store dolls and outfits.

Inside Back Cover:

Perhaps in a store dressing room, a large three-way mirror reflects a red chair, blue rug, and plant. The cover has a carry-pocket to store dolls and outfits.

Back Cover:

Bubble Barbie, Skipper and Skooter walk together carrying many packages.

#1976 *Barbie, Skipper, Skooter* Paper Dolls

YEAR	NUMBER	PRICE	TITLE	MAKER	VALUE
1966	#1980	59¢	Meet Francie Barbie's 'MOD'ern Cousin Paper Dolls	Whitman	$85.00

An excellent paper doll book because of the Francie photographs on the cover and the complete series of original outfits from #1250 to #1261.

Cover:

Three photographs of Francie #1140 dolls with one brunette Francie wearing #1257 Dance Party, a second brunette Francie wearing #1255 Polka Dots 'n Raindrops and a blonde Francie dressed in #1250 Gad-Abouts.

Paper Dolls:

Two punch-out dolls: Straight leg blonde Francie #1140 wears the original Francie doll two-piece white swimsuit with red dots on the top and red checks on the bottom. Bendable leg brunette Francie #1130 wears the original Francie doll one-piece swimsuit with red and green blocks and daisies.

Outfits:

Six pages of punch-out outfits.
#1250 Gad-Abouts
#1251 It's a Date
#1252 First Things First
#1253 Tuckered Out
#1254 Fresh As a Daisy
#1255 Polka Dots 'n Raindrops
#1256 Concert in the Park
#1257 Dance Party
#1258 Clam Diggers
#1259 Check-Mates
#1260 First Formal
#1261 Shoppin' Spree

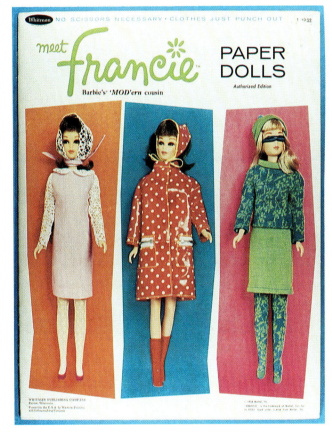

#1980 *Meet Francie Barbie's 'MOD'ern Cousin Paper Dolls.*

Accessories:

Hats, purses, flowers, scarf, and shawl.

Inside Front Cover:

Four Francie pictures: one with a long flip hair-style, two wearing hats, and one with a hair bow. The cover has a carry-pocket to store dolls and outfits.

Inside Back Cover:

Four close-up Francie pictures with a long flip hair-style, wearing a hat, a head scarf, and a headband. The cover has a carry-pocket to store dolls and outfits.

Back cover:

Five small Francie dolls are dressed in original Francie doll outfits: #1251 It's a Date, #1257 Dance Party, #1256 Concert in the Park and #1259 Check-Mates. (One outfit is unknown and not original.)

YEAR	NUMBER	PRICE	TITLE	MAKER	VALUE
1967	#1976	59¢	Barbie Has A New Look! Paper Dolls	Whitman	$90.00

An exceptional collection of #1600 doll outfits — especially the beautiful #1695 Evening Enchantment and #1697 Formal Occasion. Colorful and detailed roses decorate each page.

Cover:

A close-up of Twist 'n Turn Barbie #1160 holding a party invitation. Barbie is dressed in #1690 Studio Tour while to her right, two Twist 'n Turn Barbie dolls model the new "mod" fashions, #1687 Caribbean Cruise and #1690 Studio Tour, with new long hairdos.

Paper Dolls:

Two press-out dolls: Twist 'n Turn Barbie #1160 paper dolls have waist length straight hair, bangs, and long dangling earrings. The blonde Twist 'n Turn Barbie wears a strapless flowered swimsuit, while the brunette Twist 'n Turn Barbie wears #1685 Underprints fabric bra and pants.

Outfits:

Six pages of press-out outfits.
#1440 Pink Sparkle (white dress with pink)
#1460 Tropicana
#1470 Intrique
#1653 International Fair
    (skirt has red, white and blue vertical pleats)
#1654 Reception Line (aqua)
#1668 Riding in the Park
#1675 Sunday Visit (gold)

#1976 *Barbie Has a New Look! Paper Dolls*

Outfits (continued):

#1683 Sunflower
#1686 Print Aplenty
#1687 Caribbean Cruise
#1688 Travel Togethers
#1690 Studio Tour
#1691 Fashion Shiner
#1692 Patio Party
#1695 Evening Enchantment
#1697 Formal Occasion (cape is light pink with hood and lining in deep pink)

Accessories:

Hats and purses.

Inside Front Cover:

In her family room, Barbie wears #1687 Caribbean Cruise and uses a feather duster on her coffee table. The cover has a carry-pocket to store dolls and outfits.

Inside Back Cover:

In the kitchen, Barbie puts a cake in the oven. The cover has a carry-pocket to store dolls and outfits.

Back Cover:

Preparing for a party, Barbie dressed in #1692 Patio Party hangs paper lanterns.

Paper doll outfits #1644 On the Avenue, #1695 Evening Enchantment, and #1690 Studio Tour from #1976 *Barbie Has A New Look!* book.

YEAR	NUMBER	PRICE	TITLE	MAKER	VALUE
1968	#1976	59¢	Barbie, Christie, Stacey Paper Dolls	Whitman	$80.00

The Stacey paper doll and Stacey on the cover have blonde hair while the back inside cover and back cover picture Stacey as a redhead. The book is a good study of the less known original doll outfits.

Cover:

Three paper dolls, blonde Talking Barbie #1115, Talking Christie #1126, and a blonde Talking Stacey #1125 on stands with each doll having one hand on her hip.

Paper Dolls:

Three press-out dolls: Talking Barbie #1115 in one-piece pink swimsuit with large dark pink dots. Talking Christie #1126 in a strapless swimsuit in shades of green and royal blue stripes. Talking Stacey #1125 in a two-piece diagonal multi-stripe swimsuit in pink, blue, white, and green stripes similar to the original Talking Stacey outfit.

Outfits:

Six pages of press-out fashions.

Barbie and Stacey Paks

#1581 Extra Value Ensemble Now Wow
#1581 Extra Value Ensemble Twinkle Togs
#1582 Extra Value Ensemble Smasheroo
#1582 Extra Value Ensemble Dreamy Pink
#1582 Extra Value Ensemble Tunic 'n Tights
#1582 Extra Value Ensemble Fancy-Dancy

World of Barbie and Stacey Fashions

#1804 Knit Hit
#1813 Snug Fuzz
#1814 Sparkle Squares (pink, orange, and green)
#1822 Swirly-Cue
#1823 Jump into Lace (no frills at neck)
#1824 Snap-Dash (yellow without collar)
#1842 Togetherness
#1843 Dancing Stripes (blue horizontal stripe added)
#1844 Extravaganza

#1976 *Barbie, Christie, Stacey Paper Dolls.*

#1845 Scene-Stealers
#1846 Trail-Blazers (with blue)
#1848 All that Jazz
    (deep pink, orange, and green)

Accessories:

Purses, hat, and boots.

Inside Front Cover:

Close-ups of Talking Barbie, Christie and Stacey dolls with microphones. Printed above them are their statements, "Let's have a costume party, I love to try on clothes, Would you like to go shopping?, Being a model is terribly exciting!, I think miniskirts are smashing!, Should I change my hair style?, What shall I wear to the prom?, Would you like to have a fashion show?" The cover has a carry-pocket to store dolls and outfits.

Inside Back Cover:

Three Talking dolls walk together. Barbie wears #1814 Sparkle Squares, while Christie wears #1846 Trail-Blazers, Stacey wears #1842 Togetherness. The cover has a carry-pocket to store dolls and outfits.

Back Cover:

Models Barbie, Christie and Stacey pose in different ways with Stacey dressed in #1210 Hill-Riders.

YEAR	NUMBER	PRICE	TITLE	MAKER	VALUE
1968	#1978	69¢	Barbie, Christie, Stacey	Whitman	$80.00

The same as #1976 except for the different stock number and price change.

YEAR	NUMBER	PRICE	TITLE	MAKER	VALUE
1968	#1991	49¢	Tutti Paper Dolls	Whitman	$80.00

Press-out pieces and set-up Tutti doll's bedroom. Only the paper dolls and cut outfits are available. Because of cute little Tutti and all the original outfits, the book is fantastic.

Cover:
Tutti #3550 in the original doll outfit holds a watering can for sprinkling flowers.

Paper Dolls:
Two press-out dolls: Blonde Tutti #3555 wears a playsuit in a pink print and brunette Chris #3570 wears a blue print playsuit.

Outfits:
Unknown number of pages.
#3550 Tutti doll original outfit
#3580 Tutti doll original outfit
#3553 Night-Night Sleep Tight doll original outfit (with doll)
#3554 Me and My Dog original doll outfit
#3555 Melody in Pink original doll outfit
#3556 Sundae Treat doll original outfit
#3601 Puddle Jumpers
#3602 Ship-Shape
#3603 Sand Castles
#3604 Skippin' Rope
#3607 Come to My Party
#3609 Plantin' Posies

#1991 *Tutti and Chris Paper Dolls.*

Cut Tutti and Chris paper doll outfits from #1991 book.

YEAR	NUMBER	PRICE	TITLE	MAKER	VALUE
1969	#1976	59¢	**Barbie Dolls and Clothes**	Whitman	$65.00

A beautiful cover and many wonderful original doll outfits; however the hair color of the paper dolls, ash-blonde with green streaks, is unattractive.

Cover:

A beautiful close-up drawing of Talking Barbie #1115.

Paper Dolls:

Two press-out dolls: Blonde Talking Barbie #1115 wears a flowered, strapless swimsuit and a blonde Standard Barbie #1190 wears a one-piece black swimsuit with white and gold flowers.

Outfits:

Six pages of press-out outfits.

#1128 Julia doll original jumpsuit

#1476 Dream Wrap

#1478 Shift into Knit

#1479 Leisure Leopard

#1481 Firelights

#1482 Important In-Vestment

#1483 Little Bow-Pink

#1484 Yellow-Mellow

#1486 Winter Wow (red)

#1487 Shirtdressy

#1488 Velvet Venture
    (red dress and dark green coat)

#1489 Cloud 9

#1491 Red White 'n Warm

#1492 Silver Polish (no coat)

#1493 Fab Fur

#1865 Glo-Go

#1866 Movie Groovie

#1868 Happy Go Pink

#1869 Midi Magic

#1870 Midi Marvelous (orchid)

#1871 Romantic Ruffles (pink skirt with white dots)

#1873 Plush Pony

#1976 *Barbie Dolls and Clothes.*

Accessories: Purses.

Inside Front Cover:

Many tall flowers and leaves are drawn across a pink background. The cover has a carry-pocket to store dolls and outfits.

Inside Back Cover:

Same design as inside front cover. This cover also has a carry-pocket to store dolls and outfits.

Back Cover:

Talking Barbie and Standard Barbie stand in the center of a flower-trimmed border. Talking Barbie wears #1871 Romantic Ruffles and Standard Barbie wears #1870 Midi Marvelous.

Paper Doll Books of the 1970s

The original Barbie and family dolls experienced dramatic changes in the 1970s which were reflected in the paper doll books. Barbie abandoned glamor for the "mod" look with long, straight hair and "far out" fashions, such as bold colored miniskirts, denim jeans, bell bottom trousers, halter tops, and maxi coats, often trimmed with fringe and worn over high, lace-up boots. Barbie Boutique, Quick Curl, Malibu, SuperStar, Sun Valley, Sweet 16, and Free Moving original dolls were represented in books. However, many paper doll outfits of this time are artists' original designs or are replicas of unknown original outfits. Ken doll's and P.J. doll's outfits were also "mod" and colorful, with P.J. wearing oversized colored sunglasses. Other paper doll family members of this decade include Skipper, Christie, Francie, Stacey, Curtis, Cara, and Kelley. In many books, the paper dolls appear very young, such as the six paper dolls in #1987 *The World of Barbie*. By 1977, Barbie returned to her glamorous ways with the introduction of #1983 SuperStar.

Several books from the 1970s are noteworthy because of their innovative designs. The #1974 *Groovy P.J.*, #1975 *Pose 'n' Barbie* and #1960 *Hi! I'm Skipper* are each shaped in the outline of the cover doll, rather than the standard rectangle. The #1981 *P.J. Cover Girl* has a paper doll which is sitting rather than in the traditional standing pose and #1982 *Francie Growin' Pretty Hair* has a slit in the body of the paper doll for inserting a different head and changing her hairstyle.

YEAR	NUMBER	PRICE	TITLE	MAKER	VALUE
1970	#1976	69¢	**Barbie and Ken Paper Dolls**	Whitman	$45.00

The latest '70s fashion – long hair and colorful, trendy outfits for the "mod" group. The Twist 'n Turn Barbie and Talking Ken paper dolls do not entirely resemble the original dolls because Barbie doll's hair is shorter and Ken doll's hair is parted on the other side.

Cover:

A large blonde Twist 'n Turn Barbie #1160 poses in blue bell bottom trousers, white long sleeved blouse with bowtie and red vest. A small, smiling Talking Ken #1111 stands in the background dressed in purple bell bottom trousers and an orchid jacket.

Paper Dolls:

Two press-out dolls: Blonde Twist 'n Turn Barbie #1160 wears a strapless, one-piece white swimsuit with horizontal red stripes and brunette Talking Ken #1111 is in a blue T-shirt and flowered trunks.

Outfits:

Three pages of Barbie press-out outfits picturing bell bottom trousers, jackets, vests, minidresses, and long, granny dresses. Three pages of Ken press-out outfits picturing bell bottom trousers, suits, and shorts.

Barbie, P.J., and Julia:

#1453 Flower Wower (red)

Ken and Brad Ensemble Pak:

#1430 Town Turtle (light blue)

Accessories:

Barbie: Head scarf, neck scarf, tights with attached shoes, and knee socks with attached shoes.
Ken: Socks with attached shoes

Cardboard insert:

A cut-out poster with a large yellow sun rising behind a small Talking Ken #1111 dressed in flowered bell bottom trousers and gray jacket. In the foreground, a large Twist 'n Turn Barbie wears burgundy bell bottom trousers, vest, white blouse with a flowered red and white tie scarf. The background and border on two sides have a colorful geometric pattern.

#1986 or #1976 *Barbie and Ken Paper Dolls*

Back cover:

Three Barbie dolls pose against a colorful background. Two Barbie dolls wear short dresses trimmed with large bows. The third Barbie doll wears bell bottom trousers, a sleeveless top with a metal belt, and tie scarf in her hair.

YEAR	NUMBER	PRICE	TITLE	MAKER	VALUE
1970	#1985	69¢	Barbie and Ken Paper Dolls	Whitman	$45.00

The same as #1976 except for different stock number.

YEAR	NUMBER	PRICE	TITLE	MAKER	VALUE
1970	#1986	79¢	Barbie and Ken Paper Dolls	Whitman	$45.00

The same as #1976 except for different stock number and price change.

YEAR	NUMBER	PRICE	TITLE	MAKER	VALUE
1970	#1981	69¢	New 'n' Groovy P.J. Paper Doll	Whitman	$50.00

The title of the book is appropriately "New 'n' Groovy," as the book shows the newest trends in fashion with long hair, minidresses, and tights. The unusual cover is flamboyant, yet pleasing.

Cover:

A close-up of Twist 'n Turn P.J. #1118 wearing large pink sunglasses that magnify her eyes. The cover is cut in several places to show the cardboard insert below, revealing the initials "P.J." and two small dancing Twist 'n Turn P.J. #1118 dolls.

Paper Doll:

One press-out doll: Twist 'n Turn P.J. #1118 is dressed in a blue strapless two-piece swimsuit with large green circles. The swimsuit is held together at the midriff with one circular hoop.

Outfits:

Six pages of press-out outfits including a long dress, a midi coat, bell bottom pants, and short fitted dresses with vests, buttons, and chains. None are original Barbie doll outfits.

Accessories:

Boots, knit caps, head scarf, knit poncho, cape, tights with shoes, knee socks with shoes.

Cardboard Insert:

A poster cut-out with a large seated Twist 'n Turn P.J. #1118 wearing a pink and red long-sleeved blouse, short white vest, short red and blue checked skirt, white knee socks with red and blue oxford shoes. To the right of P.J. are the large initials "P.J." and two small dancing Twist 'n Turn P.J. dolls that show through the cover.

Back Cover:

The initials "P.J." and five small posing P.J. dolls wearing outfits shown in the book.

#1981 *New 'n' Groovy P.J. Paper Doll*

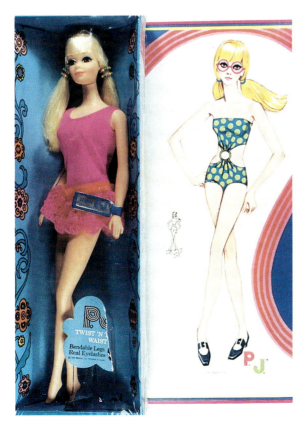

P.J. doll and paper doll from #1981 book. (Note: Doll and paper doll do not match.)

P.J. poster from #1981 book.

YEAR	NUMBER	PRICE	TITLE	MAKER	VALUE
1971	#1976	69¢	Groovy World of Barbie and Her Friends Paper Dolls	Whitman	$50.00

An excellent book of Barbie, P.J., Christie, and Stacey Paks with almost all Barbie and Francie original doll outfits. A colorful variety of groovy outfits including bell bottoms, miniskirts with fringe, and boots.

Cover:

A large Living Barbie #1116 sings into a microphone while smaller Twist 'n Turn dolls, Christie #1119, Francie #1170 and Stacey #1165 walk behind her. Barbie is dressed in #3401 Fringe Benefits, Christie in #3402 Two-Way Tiger, Francie in #3449 Buckeroo Blues and Stacey in #3408 Super Scarf.

Paper Dolls:

Four youthful press-out dolls: Living Barbie #1116 and Twist 'n Turn Stacey #1165 wear the same style one-piece multi-color swimsuit in a different design and color. Twist 'n Turn Christie #1119 has a one-piece swimsuit with a purple top, red bottom with yellow trim. Twist 'n Turn Francie #1170 has a one-piece royal blue swimsuit.

Outfits:

Six pages of press-out fashions.
#1144 Growing Pretty Hair original doll outfit
#1155 Live Action or #1152 Live Action on Stage
   original doll outfit
#1156 Live Action or #1152 Live Action on Stage
   original doll outfit

Barbie, P.J., Christie, and Steffie Paks:

#3401 Fringe Benefits	#3408 Super Scarf
#3402 Two-Way Tiger	#3411 Poncho Put-On
#3403 Baby Doll Pinks	#3413 Golfing Greats
#3404 Glowin' Out	#3414 Satin Slumber
#3407 Midi Mood	

Francie Paks:

#3444 Midi Plaid
#3445 Zig-Zag Zoom
#3449 Buckeroo Blues

Accessories:

Head scarves, knitted caps, purses, and wigs with headbands.

Inside Front Cover:

Barbie, Christie, Francie, and Stacey sing on stage in a carnival tent. Parents and children hold balloons and watch the performance. The cover has a carry-pocket for storing dolls and outfits.

#1976 Groovy World of Barbie and Her Friends Paper Dolls

Inside Back Cover:

A large red banner, large blue ribbon, many large colorful balloons, and a very large ice cream cone fly above a carnival with a ferris wheel and roller coaster. The cover has a carry-pocket for storing dolls and outfits.

Back Cover:

At the carnival, a large Living Barbie #1116 is dressed in #3401 Fringe Benefits, sings into a microphone, and waves a banner.

YEAR	NUMBER	PRICE	TITLE	MAKER	VALUE
1970	**#1981**	**69¢**	**P. J. Cover Girl Paper Doll**	**Whitman**	**$50.00**

The P.J. paper dolls and outfits cannot be identified as original paper dolls and original outfits of that time. The book shows one sitting paper doll which is rare; however, the other two paper dolls are standing in unsightly awkward positions. Although the one doll is unusual, overall the book is unattractive.

Cover:

Long-legged P.J. called Twist 'n P.J. in the book poses in denim overalls. The cover is cut to show the initials P.J. from the cardboard insert below.

Paper Dolls:

Three press-out dolls: Blonde Talking P.J. #1113, named Relax 'n' P.J., sits beneath a large red and gold sun wearing a two-piece blue, orchid, and red bikini and large sunglasses that magnify her eyes. In standing poses, brunette Twist 'n P.J. wears a one-piece swimsuit with orange, blue, and white horizontal stripes and blonde Malibu P.J. #1187 named Pose 'n P.J., wears a two-piece orange bikini.

Outfits:

Six pages of press-outfits, two pages for each P.J. doll. To match the doll with the outfit, the doll's name appears on each outfit page and on each doll's stand. The outfits featured are called, "Fun-Way Fashions, Country Styles, Hot Times Outfits, Party Go-Togethers, Sleepytime Styles and Pacesetters."

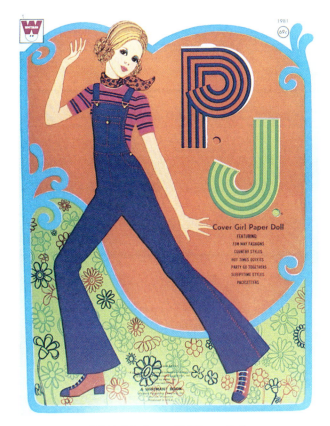

#1981 *P.J. Cover Girl Paper Doll*

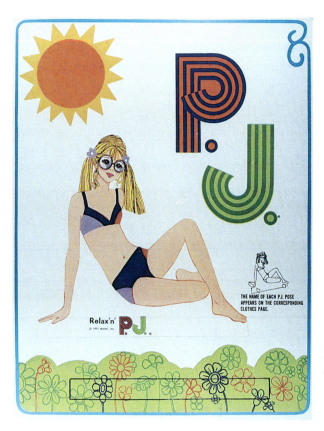

Accessories:

Head scarves, stockings with shoes and boots

Back cover:

Five small P.J.s pose in different ways.

Relax 'n P.J. paper doll in rare sitting position from #1981 *P.J. Cover Girl* book.

YEAR	NUMBER	PRICE	TITLE	MAKER	VALUE
1971	**#1987**	**69¢**	**World of Barbie Paper Dolls**	**Whitman**	**$50.00**

Each doll stands in a different position and therefore has her own clothes. A color dot on each outfit corresponds to the color dot on the doll's stand. An excellent paper doll book to study Barbie and her family of dolls and outfits of the time.

Cover:

A large group of Twist 'n Turn dolls, Barbie #1160, P.J. #1118, Stacey #1165, Casey #1180, Francie #1170 and Christie #1119, pose in a variety of ways. Barbie, Casey, and Christie model #1454 Loop Scoop, #1452 Now Knit and #1761 Sunny Slacks, respectively.

Paper Dolls:

Six youthful press-out dolls: Blonde Barbie #1160 wears the Standard Barbie #1190 swimsuit. Blonde P.J. #1118 wears a one-piece pink swimsuit with a skirt and drop waist, similar to the original swimsuit. Blonde Casey #1180 wears Talking Casey's #1125 swimsuit. Brunette Stacey #1165, blonde Francie #1170, and Christie #1119 wear their original swimsuits.

Outfits:

Six pages of press-out outfits.

Francie and Casey Paks:
#1221 Tennis Tunic
#1225 Snazz
#1227 Long On Looks
#1228 Sissy Suit
#1230 Merry Go-Rounders
#1232 Two for the Ball
#1234 The Combination

Barbie, P.J., Christie, Stacey, and Julia Paks:
#1462 Rare Pair
#1465 Lemon Kick
#1480 Pak Little Bow-Pink
#1485 Pak Shirtdressy

Barbie and Stacey Extra Value Paks
(Four outfits for each number)
#1581 Twinkle Togs
#1581 Team Ups
#1582 Tunic 'n Tights

Francie and Casey Paks:
#1761 Sunny Slacks
#1763 The Entertainer
#1862 Country Capers
#1863 Pretty Power
#1868 Happy Go Pink
#1869 Midi Magic
#1871 Romantic Ruffles
#1872 See-Worthy
#1874 Fab City (white bodice and shawl)
#1881 Made for Each Other

Accessories:
Hats, purses, and tennis racket.

Inside Front Cover:
A beautiful, large close-up of Twist 'n Turn Barbie.

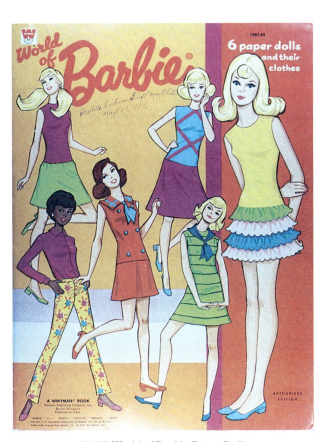

#1987 *World of Barbie Paper Dolls*

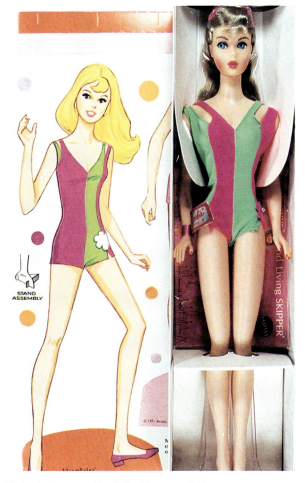

Twist 'n Turn Barbie doll with paper doll from #1987 book.

Inside Back Cover:

Twist 'n Turn Barbie mirror image of the inside front cover.

Inside Cover:

Remove the outfit pages and press out the handles to assemble a tote to carry dolls and outfits.

Back Cover:

Twist 'n Turn doll's mirror image of the cover.

YEAR	NUMBER	PRICE	TITLE	MAKER	VALUE
1972	#1974	39¢	Groovy P.J. Paper Doll Fashions	Whitman	$40.00

This particular book shape was used for only two other Barbie and family paper doll books, Pos 'n Barbie and Hi! I'm Skipper. The unusual shape, photographic cover, and colorful outfits make this book unique and highly collectible.

#1974 *Groovy P.J. Paper Doll Fashions.*

Cover:

An enlarged photograph of Live Action P.J. #1156 or Live Action on Stage P.J. #1153 holds a model's hat box. The book is cut around P.J. wearing #1057 Groovin' Gauchos.

Paper Dolls:

One press-out paper doll: Live Action P.J. #1156 or Live Action on Stage P.J. #1153 dressed in a flowered bikini.

Outfits:

Six pages of press-out outfits.

Fashion Original: #3483 Purple Pleasers, #3490 Party Lines

Best Buy: #3356 Good Sports, #3356 Sport Star, #3363 Fancy That Purple

Fashions 'n Sounds for Barbie, P.J., and Christie: #1057 Groovin' Gauchos

P.J., Christie, and Stacey: #3418 Magnificent Midi (yellow)

Accessories:

Hats, scarves, and boots

YEAR	NUMBER	PRICE	TITLE	MAKER	VALUE
1972	#1975	39¢	Pos 'n' Barbie Paper Doll Fashions	Whitman	$40.00

This particular book shape was used for only two other Barbie and family paper doll books, Groovy P.J. and Hi! I'm Skipper. The unusually shaped photographic cover and colorful outfits make this an exceptional book.

Cover:

An enlarged photograph of Talking Barbie #1115 holding a model's hat box. The book is cut around Barbie dressed in Best Buy Fashion #3342 All American Girl.

Paper Doll:

One press-out paper doll: Talking Barbie #1115 wears a two-piece blue bikini with orange trim.

Outfits:

Six pages of press-out outfits.

Best Buy Fashions: Ensemble assortment.
(Six outfits for each number.)
#3356 Picture Me Pretty
#3356 Glowin' Gold
#3363 Pleasantly Peasanty
#3363 Perfect Purple

Fashion Original: #3487 Sleepy Set, #3491 Suede 'n Fur

Francie and Casey Pak: #1769 Long On Leather (blue)

Accessories: Hats, boots, and flowers.

#1975 *Pos 'n' Barbie Paper Doll Fashions.*

YEAR	NUMBER	PRICE	TITLE	MAKER	VALUE
1972	#1994	79¢	**Malibu Barbie The Sun Set**	Whitman	$30.00

Because Malibu Barbie is a common doll and the paper doll outfits are not original Barbie doll outfits, this paper doll book has little value.

Cover:

At the seashore, Malibu dolls Francie #1068, Ken #1088, Barbie #1067, and Skipper #1069 walk arm-in-arm dressed in their original swimsuits.

Paper Dolls:

Four press-out paper dolls: Malibu Barbie #1067 is dressed in a one-piece light blue swimsuit while Francie #1068 wears the original suit with a purple top. Both dolls have long white hair streaked with pink and green, respectively. Blonde Malibu Ken #1088 wears purple trunks while blonde Skipper #1069 wears a two-piece orange swimsuit. The color of the doll's stand corresponds to the color tabs on the clothes.

Outfits:

Six pages of punch-out outfits. The fashions consist of long dresses, bell bottom trousers and jackets. None of the "Sun Fashions" (name printed on cover) are original doll outfits.

Accessories: None

Back Cover:

At the beach, four Malibu dolls — Francie #1068, Ken #1088, Barbie #1067, and Skipper #1069 — enjoy the scenery, a gigantic sun above the blue water.

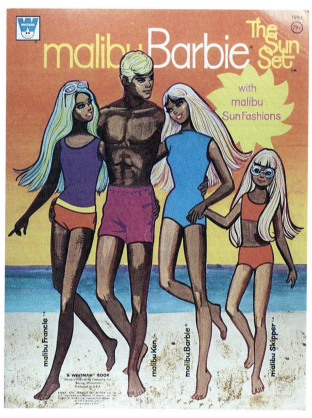

#1994 *Malibu Barbie The Sun Set*

YEAR	NUMBER	PRICE	TITLE	MAKER	VALUE
1972	#1994	79¢	**Malibu Barbie The Sun Set**	Whitman	$30.00

The same stock number and price except four instead of six outfit pages.

YEAR	NUMBER	PRICE	TITLE	MAKER	VALUE
1973	#1954	39¢	Barbie's Boutique	Whitman	$30.00

A written name describes the outfits on each page as Pink Polka Dot, Ruffles 'n Lace, Fancy in Flowers and Check Chums which are not the original outfit names. An outstanding cut-out book with an interesting theme, a smiling Barbie, and outfits with checks, dots, flowers, lace, and bows.

Cover:

A blonde Quick Curl Barbie #4220 poses in the Best Buy Fashion #3205 with a red checked skirt. To the right of Barbie are three circles with a hat, a large black and white bow, and a tan purse.

Paper Doll:

One press-out doll: Quick Curl Barbie #4220 with curly blonde hair wears a one-piece swimsuit divided into four sections, two pink sections with white lace flowers diagonally across from two black and white check sections.

Outfits:

Six pages of cut-out outfits.

Quick Curl Kelley: #4221 original doll outfit, green gingham with ruffle

Quick Curl Francie: #4222 original doll outfit, pink polka dot, no ruffle

Best Buy:
#3203 Polka Dot Perfect
#3205 Gingham 'n Lace
#3206 Simply Summer
#3208 Anytime Orange
#3346 Aboard In Blue
#3347 Short 'n Sweet
#3348 Sunny Sleep-Ins (gold)
#8620 Fancy In Flowers

Accessories: Hats, head scarves, purses, and bags.

Inside Front Cover:

Green handbag, long green gloves, green clock, pink hat, pink purse, and gold and plaid travel bag beside the Barbie paper doll.

#1954 *Barbie's Boutique Paper Doll Book.*

Inside Back Cover:

A large, pink wardrobe trimmed in pink and blue gingham has a carry-pocket for storing the doll and outfits.

Back Cover:

Barbie dressed in #3203 Polka Dot Perfect visits a boutique.

Barbie paper doll with outfits from #1954 book.

YEAR	NUMBER	PRICE	TITLE	MAKER	VALUE
1973	#1954	59¢	Barbie's Boutique	Whitman	$30.00

The same as #1954 except for a different price and no inside back cover wardrobe.

YEAR	NUMBER	PRICE	TITLE	MAKER	VALUE
1973	#1974-1	59¢	Barbie's Boutique	Whitman	$30.00

The same as #1954 except for a different stock number, price change, and without inside back cover wardrobe.

YEAR	NUMBER	PRICE	TITLE	MAKER	VALUE
1973	#1984	79¢	Quick Curl Barbie and Her Paper Doll Friends	Whitman	$30.00

Although none of the fashions are original doll outfits, the four Quick Curl dolls with the different hair styles and punch-out wigs make this a delightful book.

Cover:

Four Quick Curl doll portraits, Barbie #4220, Francie #4222, Kelley #4221, and Skipper #4223 — each with a different hairstyle. Autographs appear beneath each photograph.

Paper Dolls:

Four press-out dolls: Barbie wears a geometric print bikini while Francie, Kelley, and Skipper wear geometric print strapless one-piece swimsuits. The colored dot on each doll's stand corresponds to the color tabs on the clothes.

Outfits:

Six pages of press-out outfits.

Accessories:

Eight wigs and a head scarf.

Inside Cover:

Remove the outfit pages and press out the handles to assemble a tote to carry dolls and outfits.

Back Cover:

Skipper and Kelley at the beauty parlor getting their hair styled by Barbie and Francie.

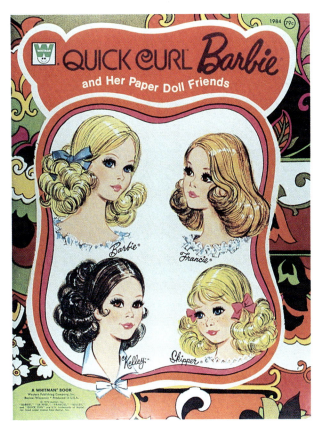

#1984 *Quick Curl Barbie and Her Paper Doll Friends*

YEAR	NUMBER	PRICE	TITLE	MAKER	VALUE
1973	#1984	79¢	Quick Curl Barbie And Her Paper Doll Friends	Whitman	$30.00

The same as #1984 with four outfit pages instead of six.

YEAR	NUMBER	PRICE	TITLE	MAKER	VALUE
1973	**#1990**	**69¢**	**Barbie Country Camper Doll Book**	**Whitman**	**$30.00**

A good reference guide to Best Buy outfits.

Cover:

A large close-up of Malibu Barbie #1067 waving to Ken and P.J. driving the Country Camper #4994 in the background. Barbie wears a one-piece swimsuit with a flowered shawl around her shoulders.

Paper Dolls:

Three press-out Malibu dolls: Malibu Barbie #1067 and Malibu P.J. #1187 wear one-piece swimsuits, pink and orchid respectively. Ken wears blue trunks.

Outfits:

Six pages of press-out outfits with two outfit pages for each doll. The Best Buy outfits are usually unnamed.

Barbie Best Buy:
#3203 (blue with pockets); #3205 (green); #3206; #3208; #3363 Pants, Perfect Purple.

Ken Best Buy: #8615; #8617; #8618 (aqua and deep pink.)

Francie Best Buy: #8644

Francie Ensemble Pak: #1241 Striped Types (pink shirt)

Fashion Originals, Francie Assortment:
#3276 The Slack Suit   #3287 Smashin' Satin
#3285 Peach Treats   #3488 Overall Denim
#3286 Double Ups

Accessories: Hats.

Cardboard Insert:

Three colorful sleeping bags have slits for Barbie, Ken, and P.J. Each sleeping bag has a doll's name.

Inside Cover:

Remove the outfit pages to reveal the Country Camper #4994 with a carry-pocket.

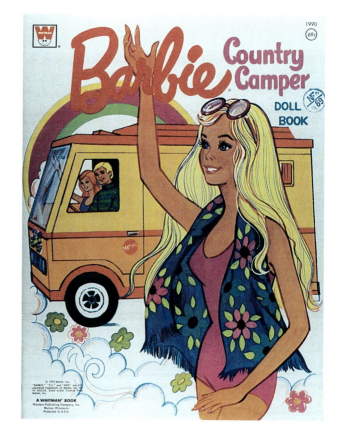

#1990 *Barbie Country Camper Doll Book*

Back Cover:

Four small country campers driving down a curving road.

YEAR	NUMBER	PRICE	TITLE	MAKER	VALUE
1973	#1996	79¢	Barbie's FriendShip Paper Dolls	Whitman	$25.00

The Barbie and Ken United Airline outfits are special.

Cover:

Accompanied by the pilot, Twist 'n Turn Barbie #1160 leaves a United Airline airplane.

Paper Dolls:

Two press-out dolls: Twist 'n Turn Barbie #1160 wears a one-piece strapless swimsuit with wide white and pink diagonal stripes and narrow purple and green diagonal stripes. Talking Busy Ken #1196 wears dark blue trunks with white stars.

Outfits:

Four pages of press-out outfits.

Ken Best Buy : #8616 (purple); #8618 (light blue).

Get Ups 'n Go: #7707 United Airline Pilot (blue).

Barbie: #7703 United Airline Stewardess.

Accessories:

Hats, purses, and suitcase.

Cardboard Insert:

Press-out Barbie's serving cart, pitcher, and three glasses.

Back Cover:

Barbie and pilot wave from the plane's ramp surrounded by posters promoting United Airlines, Hawaii, and Washington, D.C.

#1996 Barbie's FriendShip Paper Dolls.

YEAR	NUMBER	PRICE	TITLE	MAKER	VALUE
1973	#1955	39¢	**Malibu Francie Doll Book**	Whitman	$25.00

A valuable cut-out book for Francie doll collectors.

Cover:

A radiant sun smiles at a large blonde Malibu Francie #1068. Francie wears a gold halter top with large red hearts, white shorts with gold cuffs, a gold belt, and a red heart on her hip.

Paper Doll:

One press-out doll: A youthful Malibu Francie #1068 with hair streaked with green in Malibu Francie original doll swimsuit.

Outfits:

Six pages of cut-out outfits. Some outfits are named, Malibu Casuals, Flight Suit, Fun Fun Fashions, and Francie's Fun Fashions.

Accessories:

Hats, purses, boots, and belt.

Inside Back Cover:

Four Francies in different scenes: Francie giving a flower to another Francie, Francie eating an apple and Francie holding scissors. The cover has a carry-pocket for storing dolls and outfits.

Back Cover:

Malibu Francie #1068 walks through a field carrying a bouquet of flowers.

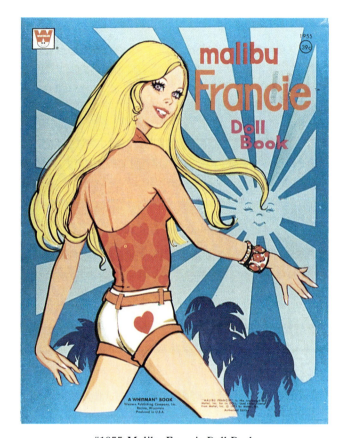

#1955 *Malibu Francie Doll Book.*

YEAR	NUMBER	PRICE	TITLE	MAKER	VALUE
1973	#1982	79¢	Francie With Growin' Pretty Hair	Whitman	$40.00

The only Barbie book with a slit in the body to accommodate a different head. Very interesting and collectible paper doll book.

Cover:

Growin' Pretty Hair Francie #1074 sits on a stool dressed in With-It-Whites #3449. Francie points with outstretched arms to two wigs on each side of her.

Paper Doll:

One press-out doll: Blonde Growin' Pretty Hair Francie #1074 has a slit above the top of her red and white dotted bikini to allow inserting a different head.

Outfits:

Four pages of press-out outfits.

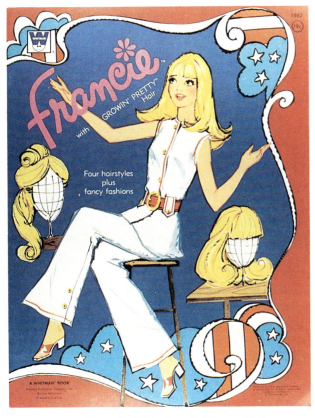

#1982 *Francie With Growin' Pretty Hair.*

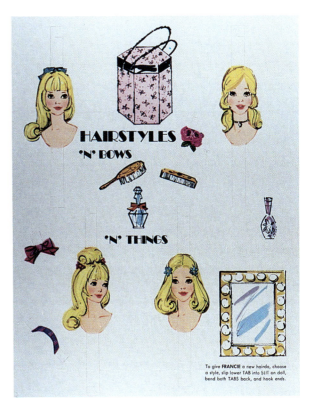

Cardboard insert with changeable heads to insert into paper doll from #1982 book.

Outfits (continued):
Fashion Original: #3276 The Slack Suit

Francie Ensemble Pak: #3448 With-It-Whites
#3461 Peach Plush

Accessories:
Hats, purses, and boots.

Cardboard Insert:
Press out a bow, hairband, and four blonde Francie heads, each with a different hair style, to change the paper doll's head.

Back Cover:
Three large Francies sitting beneath the sign, "A Model You," which lists a model's attributes.

YEAR	NUMBER	PRICE	TITLE	MAKER	VALUE
1973	#1982	79¢	Francie With Growin' Pretty Hair	Whitman	$40.00

The same as #1982 except as a book, not a folder.

YEAR	NUMBER	PRICE	TITLE	MAKER	VALUE
1973	#1969	39¢	Hi! I'm Skipper Paper Doll Fashions	Whitman	$30.00

This particular book shape was used for only two other Barbie and family paper doll books, Pos'n' Barbie and Groovy P.J. Although the Skipper paper doll is poor, the unusual shape, photographic cover, and colorful outfits make this an exceptional paper doll book.

Cover:
A large photograph of Pose 'n Play Skipper #1179 looking at a gumball machine.

Paper doll:
One press-out doll: Pose 'n Play Skipper #1179 dressed in a one-piece belted maroon swimsuit. The paper doll does not resemble Skipper.

Outfits:

Six pages of Skipper press-out outfits called "zippy and spiffy styles."

Fashion Original:

#3297 Party Pair
#3291 Nifty Knickers (green top)
#3292 Play Pants (light blue)
#3293 Dream Ins
#3296 Red White 'n Blues

Best Buy Skipper Ensemble:

Fun Runners
Flower Power
White, Bright 'n Sparkling

Accessories:

Boots, caps, scarf, vest, and stockings with shoes

#1969 *Hi! I'm Skipper Paper Doll Fashions*

YEAR	NUMBER	PRICE	TITLE	MAKER	VALUE
1973	#1952	49¢	**Malibu Skipper**	Whitman	$25.00

The large Skipper paper doll and outfits are wonderful.

Cover:

Malibu Skipper #1069 aboard a sailboat, holds the rope governing the sails. Skipper is dressed in jeans, T-shirt and life jacket.

Paper Dolls:

One press-out doll: An unusually large Malibu Skipper #1069 dressed in cut-off blue jeans and a dark blue T-shirt with horizontal red and white stripes.

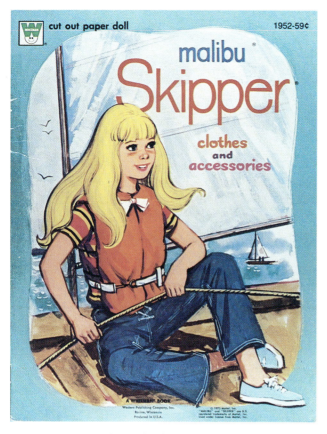

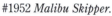

#1952 *Malibu Skipper.*

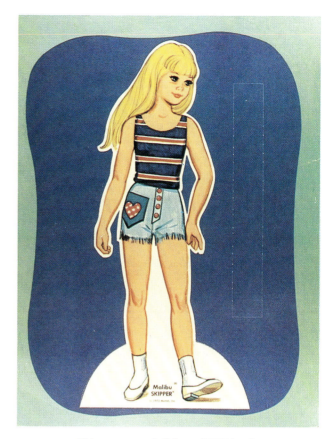

Skipper paper doll from #1952 book.

Outfits:

Six pages of Skipper cut-out outfits.

Accessories:

Hats and purses.

Inside back cover:

Skipper sailing with two boys. The cover has a carry-pocket for storing the doll and clothes.

Back cover:

At sunset, Malibu Skipper ties the sailboat to the dock.

YEAR	NUMBER	PRICE	TITLE	MAKER	VALUE
1973	#1952	59¢	**Malibu Skipper**	**Whitman**	$25.00

The same as #1952 except for a price change and no carry-pocket.

YEAR	NUMBER	PRICE	TITLE	MAKER	VALUE
1973	#1952	59¢	Malibu Skipper	Whitman	$25.00

The same as #1952 except for an encircled price, price change, and no carry-pocket.

YEAR	NUMBER	PRICE	TITLE	MAKER	VALUE
1973	#1945-2	59¢	Malibu Skipper	Whitman	$25.00

The same as #1952 with a stock number and price changes, and no carry-pocket.

YEAR	NUMBER	PRICE	TITLE	MAKER	VALUE
1974	#1951	49¢	Barbie Goin' Camping	Whitman	$20.00

Only one outfit is worth mentioning, Barbie doll's hooded jacket with hip boots, backpack, and fish in hand.

Cover:

Holding a camera, Malibu Barbie #1067 rests on a hilltop with Malibu P.J. #1187 who shields her eyes to look at the valley. The Sport Set Sun Valley Barbie #7806 or Newport Barbie #7807 wears orange shorts, orange shirt with white vertical stripes, orange and red knee socks, and tan boots. P.J. wears green trousers and shirt. To the side of Barbie and P.J. are cut-out hat, purse, dresses, and pants.

Paper dolls:

Two press-out dolls: Sun Valley Barbie #7806 or Newport Barbie #7807, wears a green one-piece swimsuit with pink vertical stripes. Malibu P.J. #1187 wears a one-piece pink swimsuit with orange vertical stripes.

Outfits:

Six pages of cut-out outfits. Barbie doll's clothes have green tabs and P.J. doll's clothes have purple tabs.

Accessories:

Hats and tennis racket.

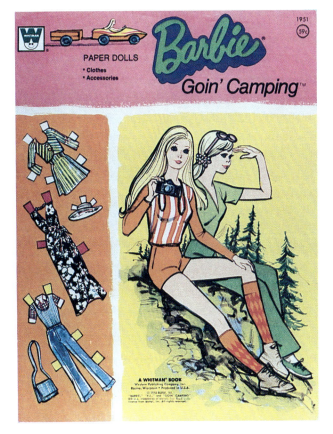

#1951 *Barbie Goin' Camping.*

YEAR	NUMBER	PRICE	TITLE	MAKER	VALUE
1974	#1951	59¢	**Barbie Goin' Camping**	Whitman	$20.00

The same as #1951 except for the price change.

YEAR	NUMBER	PRICE	TITLE	MAKER	VALUE
1974	#1981	69¢	**Barbie's Sweet 16 Paper Dolls**	Whitman	$25.00

This book is collectible because it celebrates Barbie doll's 16th birthday.

Cover:

A large close-up of Sweet 16 Barbie #7796 with two small dancing Sweet 16 Barbie dolls dressed in the original doll clothes.

Paper Dolls:

Two very youthful press-out dolls: one Sweet 16 Barbie #7796 wears an aqua and purple one-piece swimsuit, while the other Sweet 16 Barbie wears a two-piece pink swimsuit with white circles.

Outfits:

Six pages of Barbie press-out outfits.

Accessories:

Hats, purses, sunglasses, umbrella, and bouquet.

Cardboard Insert:

A colorful poster with Sweet 16 Barbie wearing purple bell bottom trousers with a frilly, sleeveless blouse and standing next to a large number "16."

Back Cover:

Three Sweet 16 Barbie dolls in different positions.

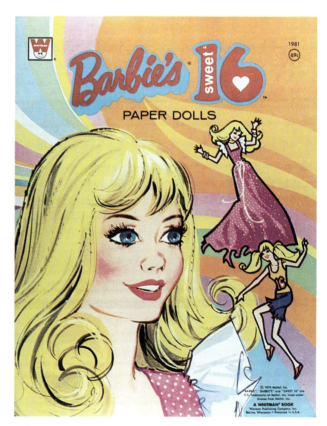

#1981 *Barbie's Sweet 16 Paper Dolls.*

YEAR	NUMBER	PRICE	TITLE	MAKER	VALUE
1975	#1956	59¢	**Yellowstone Kelley Paper Dolls**	Whitman	$15.00

A pretty outdoor cover with average paper doll outfits.

Cover:

Yellowstone Kelley #7808 dressed in the original Kelley doll outfit, blue and white vertical striped shorts and red halter with white dots, feeds a fawn. To the side, three circles picture Kelley in a mountain climbing scene, at Old Faithful, and at the camp stove.

Paper Dolls:

One press-out doll: Golden hair Yellowstone Kelley doll #7808 wears the original Kelley doll outfit.

Outfits:

Six pages of cut-out outfits.

Accessories:

Hats, purse, umbrella, head scarf, grill, eggs in a frying pan, and mug.

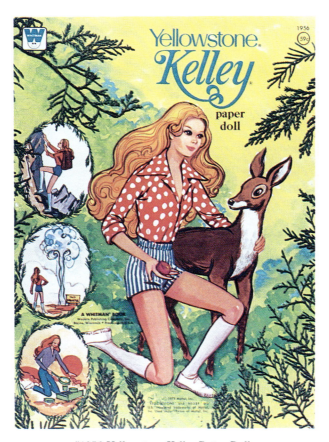

#1956 Yellowstone Kelley Paper Dolls.

YEAR	NUMBER	PRICE	TITLE	MAKER	VALUE
1975	#1981	79¢	Barbie and Her Friends All Sports Tournament	Whitman	$15.00

Cara and Curtis are welcome additions to the paper doll books. The coordinating outfits are great.

Cover:

Free Moving dolls, Barbie #7170, Ken #7280, Cara #7283, and Curtis #7282, are dressed in royal blue personalized sweatshirts.

Paper Dolls:

Four press-out dolls: Free Moving Barbie #7270 wears a one-piece red swimsuit, Ken #7280 wears aqua swim trunks, Curtis #7283 wears gold swim trunks, and Cara #7282 wears a one-piece pink swimsuit. Each swimsuit is trimmed with white.

Outfits:

Four pages of press-out outfits: skiing, tennis, golf, and sweatsuits.

Accessories:

Hats and purses.

Back cover:

Four circles picture Barbie and Ken playing tennis, Barbie and Ken skiing, Cara and Curtis playing golf, and Barbie, Ken, Curtis, and Cara at the pool.

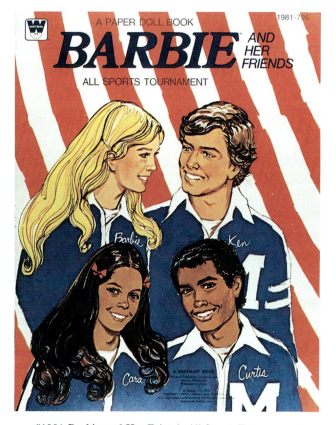

#1981 *Barbie and Her Friends All Sports Tournament.*

YEAR	NUMBER	PRICE	TITLE	MAKER	VALUE
1976	#1989	79¢	Barbie Fashion Originals Paper Doll	Whitman	$20.00

The title of the book is the same as a group of Barbie original outfits. The cover and outfits feature Barbie, the model.

Cover:

A large Malibu Barbie #1067 models her long, red and white gown with four smaller Malibu Barbies in the background also modeling their outfits.

Paper Dolls:

One press-out doll: Malibu Barbie #1067 in a colorful, flowered, one-piece strapless swimsuit.

Outfits:

Four pages of press-out outfits.

Accessories:

Hats and head scarf.

Cardboard Insert:

Press-out and assemble a colorful tote to carry the outfits. The tote is too small for the doll.

Back Cover:

Three Malibu Barbie dolls modeling their outfits.

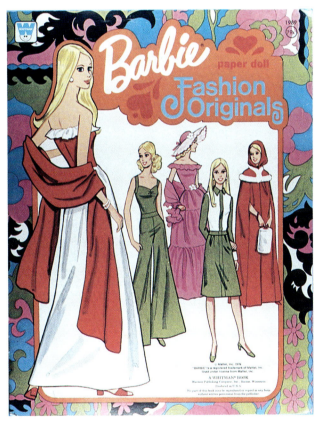

#1989 *Barbie Fashion Originals Paper Doll.*

YEAR	NUMBER	PRICE	TITLE	MAKER	VALUE
1976	#1990	79¢	**Growing Up Skipper Paper Dolls**	**Whitman**	**$20.00**

The two Skipper dolls with matching outfits make this a very special book.

Cover:

Two Growing Up Skipper #7259 dolls of different heights wear the original tall and short doll dresses.

Paper Dolls:

Two press-out dolls: The shorter Growing Up Skipper #7259 wears a red and black one-piece geometric print swimsuit. The taller Skipper wears a one-piece blue swimsuit with yellow flowers.

Outfits:

Four pages of matching press-out outfits: two #7259 Growing Up Skipper Doll outfits.

Accessories:

Hats and purses.

Cardboard Insert:

Press out and assemble a colorful flowered tote to carry outfits. The tote is too small for the dolls.

Back Cover:

Three Skipper scenes, Skipper with a bicycle, a Skipper portrait, and Growing Up Skipper in the original long dress.

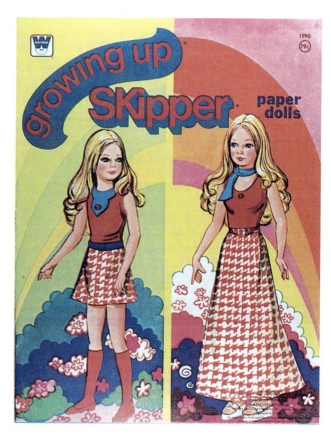

#1990 *Growing Up Skipper Paper Dolls.*

YEAR	NUMBER	PRICE	TITLE	MAKER	VALUE
1976	#1996	79¢	Barbie's Beach Bus Paper Dolls	Whitman	$15.00

The Barbie paper doll wears flat shoes and appears very young. Barbie looks like Francie.

Cover:

Malibu dolls Barbie #1067, Ken #1088, and Skipper #1069 build a sand castle at the beach. The Beach Bus #7805 is shown in the background.

Paper Dolls:

Three press-out dolls, surfboard, and three beach balls: Malibu Barbie #1067 wears a two-piece blue swimsuit, Ken #1088 wears orange trunks and Skipper #1069 wears a two-piece orange swimsuit.

Outfits:

Four pages of press-out outfits. Each doll has one named outfit page and one page has outfits for all three dolls.

Accessories: Head scarf.

Back cover:

Beach Bus #7805 with stand to punch out

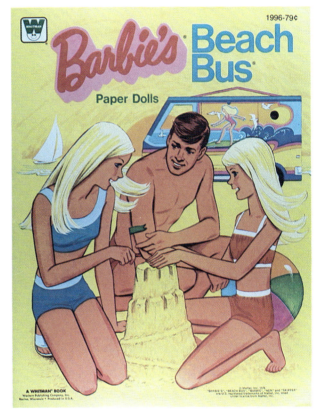

#1996 *Barbie's Beach Bus Paper Dolls.*

YEAR	NUMBER	PRICE	TITLE	MAKER	VALUE
1976	#1996-1	89¢	Barbie's Beach Bus Paper Dolls	Whitman	$15.00

The same as #1996 except for the stock number and price changes.

YEAR	NUMBER	PRICE	TITLE	MAKER	VALUE
1977	#1983	79¢	SuperStar Barbie Paper Doll	Whitman	$20.00

The cover shows a radiant, smiling SuperStar Barbie with long, flowing blonde hair. The outfits appropriately represent the original elegant SuperStar doll outfits.

Cover:

A large close-up of SuperStar Barbie #9720 dressed in the original doll outfit with a sparkling necklace and earrings. A small SuperStar Barbie poses in the background in the same gown.

Paper Doll:

One press-out doll: SuperStar Barbie #9720 wears a pink bikini with one shoulder strap.

Outfits:

Four pages of press-out SuperStar Barbie original outfits:

Best Buy: #9621; #9622; #9624 (green, yellow and pink); #9627; #9958; #9959; #9960; #9968 (green boots with red laces); #9969.

SuperStar Outfits: #9835; #9836 (light green); #9837.

Get Ups 'n Go: #9741 (dull fabric).

Accessories: None.

Cardboard Insert:

Press out and assemble a pink stage with large yellow stars and geometric design.

Back Cover:

On a pink background, three stars picture SuperStar Barbie in different poses surrounded by smaller yellow and orange stars.

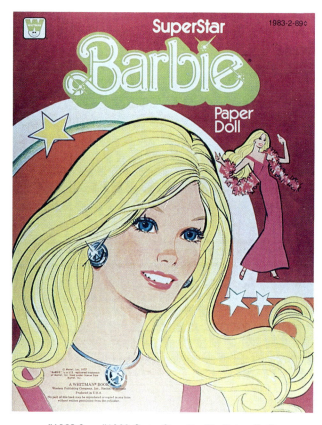

#1983-2 or #1983 *SuperStar Barbie Paper Doll.*

YEAR	NUMBER	PRICE	TITLE	MAKER	VALUE
1977	#1983-2	89¢	SuperStar Barbie Paper Doll	Whitman	$20.00

The same as #1983 except for the stock number and price change.

YEAR	NUMBER	PRICE	TITLE	MAKER	VALUE
1977	**#1993-1**	**89¢**	**Ballerina Barbie Paper Dolls**	**Whitman**	**$25.00**

Outstanding collection of ballerina outfits.

Cover:

Ballerina Barbie #9093 leaps through the air holding a bouquet. Barbie wears the original white Ballerina tutu trimmed in gold.

Paper Dolls:

Two press-out dolls: Ballerina Barbie #9093 has blonde hair streaked with gold. One ballerina stands on her toes wearing a strapless purple and gold print swimsuit, the second ballerina wears a flowered strapless swimsuit.

Outfits:

Four pages of press-out outfits.
#9093 Ballerina Barbie doll original outfit
#9326 Tchaikovsky Nutcracker Suite's
    Sugar Plum Fairy
#9327 Tchaikovsky Sleeping Beauty's
    Princess Aurora
#9328 Tchaikovsky Nutcracker Suite's
    Snow Fairy

Accessories:

Hat, purse, hair pieces, wand, and crown.

Back Cover:

A large blue diamond shape holds a portrait of Ballerina Barbie with her gold crown. Above and beneath Barbie are two smaller Ballerina Barbies in ballet positions dressed in Tchaikovsky's Sleeping Beauty and Sugar Plum Fairy outfits.

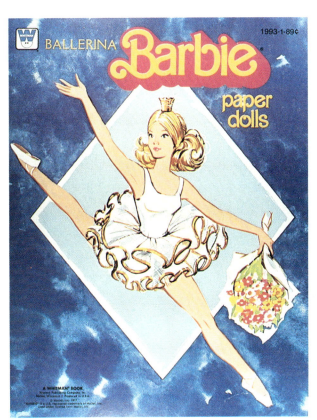

#1993-1 *Ballerina Barbie Paper Dolls.*

#9326 Nutcracker Suite's Sugar Plum Fairy outfit with matching paper doll outfit from #1993-1 book.

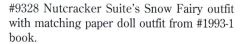

#9328 Nutcracker Suite's Snow Fairy outfit with matching paper doll outfit from #1993-1 book.

#9327 Sleeping Beauty's Princess Aurora outfit with matching paper doll outfit from #1993-1 book.

YEAR	NUMBER	PRICE	TITLE	MAKER	VALUE
1978	#1997-1	89¢	Fashion Photo Barbie and P.J.	Whitman	$15.00

A diverse wardrobe for the glamorous Fashion Photo Barbie and P.J.

Cover:

Fashion Photo Barbie #2210 poses before her camera dressed in the original Fashion Photo gown. To her side, Barbie appears in three frames of movie film.

Paper Dolls:

Two press-out dolls: Blonde Fashion Photo Barbie #2210 wears a mulit-color one-piece swimsuit while the brunette Fashion Photo P.J. #2323 is dressed in a two-piece gold swimsuit with many colored stars.

Outfits:

Four pages of press-out outfits: #2210 Fashion Photo Barbie doll original outfit.

Accessories: Purses.

Cardboard Insert:

Punch out and assemble a stage with a large, orange star in the center.

Back Cover:

A large Fashion Photo Barbie and two smaller Fashion Photo Barbie and P.J. pose in front of the camera lights.

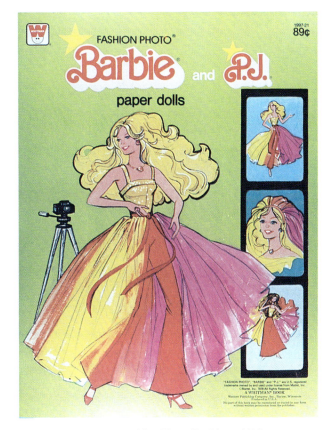

#1997-21 *Fashion Photo Barbie and P.J.*

YEAR	NUMBER	PRICE	TITLE	MAKER	VALUE
1978	#1997-21	89¢	Fashion Photo Barbie and P.J.	Whitman	$15.00

The same as #1997-1 except for the stock number.

YEAR	NUMBER	PRICE	TITLE	MAKER	VALUE
1978	#1982-32	99¢	Fashion Photo Barbie and P.J.	Whitman	$15.00

The same as #1997-1 except for the stock number and price changes.

Paper Doll Books of the 1980s

Many of the paper doll books issued in the 1980s represent the original Barbie dolls and outfits available through the decade. Original dolls and outfits appeared both on the book covers and as paper dolls. Paper doll books #1521 *Day to Night,* #1537-2 *SuperStar,* #1500 *Perfume Pretty* and #1537 *Jewel Secrets* contain only original Barbie doll outfits. #1982-47 *Barbie Fantasy* pictures many original Barbie outfits from 1983, while #1983-48 *Crystal Barbie* shows Barbie doll's original fashions from 1984 and #1982-45 *Angel Face* has all the First Barbie original outfits.

Barbie continued to reflect popular trends during the 1980s and the paper doll books followed suit. Barbie took up Western style in #1982-43 *Western,* followed the exercise craze in #1522 *Great Shape,* and was an executive in #1521 *Day-to-Night.*

Most books at this time feature only Barbie, but she appears occasionally with Ken and Skipper, and appears with Diva, Dee Dee, Dana, and Derek in #1528 *Rockers* and with Whitney in #1523 *Tropical Barbie.*

Two unusual books during this time were playbooks, #1836 *Campsite at Lucky Lake* with Skipper and #1836-43 *Pink and Pretty Playbook* which included beautiful scenery and an assortment of play accessories.

Perhaps the most outstanding book of the decade is #1527 *Barbie and Ken* which features four sets of matching formal wear for Barbie and Ken.

YEAR	NUMBER	PRICE	TITLE	MAKER	VALUE
1980	#1836	$2.00	**Barbie and Skipper** **Campsite at Lucky Lake Playbook**	**Whitman**	$20.00

The first Barbie Playbook features beautiful, smiling paper dolls with primarily sports fashions. Although the outfits are not original doll outfits, the fashions are interesting because many Barbie outfits have accessories attached like a soda bottle, hot dog, candle, book, bucket, and bundle of logs. Skipper doll's outfits have a toothbrush, toothpaste, hot dog, soda bottle, and fishing pole. Both Barbie and Skipper outfits have canoe paddles, tennis rackets, and hot dogs to cook over the campfire. The attached accessories and close attention to the book's theme make this an interesting book.

Cover:

Sun Lovin' Malibu Barbie #1067 and Sun Lovin' Malibu Skipper #1069 hike from their campsite at Lucky Lake. Three small square pictures at the bottom of the cover show items in the book, a fold-out campsite, camp furniture, and punch-out dolls and clothes.

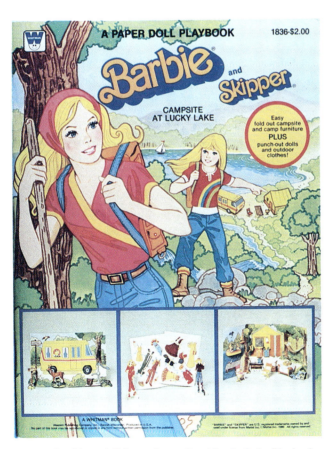

#1836 Barbie and Skipper Campsite at Lucky Lake Playbook.

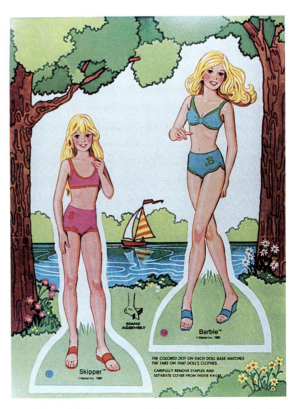

Barbie and Skipper paper dolls from #1836 playbook.

Paper Dolls:

Two punch-out dolls: Sun Lovin' Malibu Barbie #1067 and Sun Lovin' Malibu Skipper #1069 in original Sun Lovin' doll swimsuits.

Outfits:

Four pages of punch-out outfits.

Accessories: Hats

Inside Cover:

The cover opens into three parts showing a campground in the forest near Lucky Lake. A campfire burns in front of a large yellow and red tent. In the cover are slits to insert the punch-out pieces from the cardboard inserts.

Cardboard Inserts:

Three pages of punch-out pieces to insert into the campsite scene. The first page has a reindeer, owl, two ducks, waterfall, tree, hammock, and camping chair. The second page has a camping chair, backpack, bucket, squirrel, fish, skunk, bird, rabbit, small sailboat, pot, greens, plate, and Barbie and Skipper in a canoe. The third insert has a grill, grill logs, pot, backpack, bird, picnic basket, picnic table, lantern, and sign reading "Hiking Trail."

Back Cover:

Star Traveler Motor Home #9794

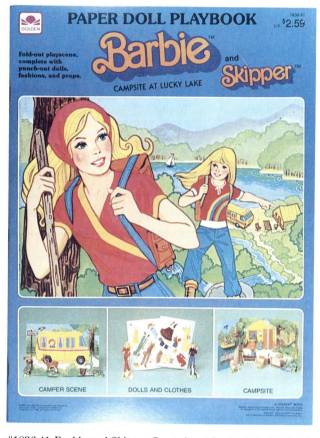

#1836-41 *Barbie and Skipper Campsite at Lucky Lake Playbook.*

YEAR	NUMBER	PRICE	TITLE	MAKER	VALUE
1980	#1836-31	$2.59	**Barbie and Skipper Campsite at Lucky Lake Playbook**	Golden	$20.00

The same as #1836 except Golden issued the book with a different stock number and price.

YEAR	NUMBER	PRICE	TITLE	MAKER	VALUE
1980	#1836-41	$2.59	**Barbie and Skipper Campsite at Lucky Lake Playbook**	Golden	$20.00

The same as #1836 except Golden issued the book in a different shade of blue background, with a different stock number and price.

YEAR	NUMBER	PRICE	TITLE	MAKER	VALUE
1980	#1980-3	89¢	Super Teen Skipper Paper Doll	Whitman	$15.00

A very pretty Skipper paper doll; however, her wardrobe is plain and too grown-up.

Cover:

A large close-up of smiling Super Teen Skipper #2756 with two small Super Teen Skippers to her side. One small Skipper rides a skateboard wearing a helmet, knee, and elbow pads; the other small Skipper wears a shirt and long skirt outfit while holding a hair brush and looking into a mirror.

Paper Dolls:

One punch-out doll: Super Teen Skipper #2756 dressed in the original doll outfit.

Outfits:

Four pages of punch-out outfits; most are not original Skipper doll outfits.

Super Teen Skipper #2756 (long skirt)

Accessories: Hats.

Cardboard Insert:

Punch out and assemble a tote to carry outfits. The paper doll is too large for the tote. Some clothes need to be folded to store in the tote.

Back Cover:

One Skipper rides on a skateboard while another Skipper holds her hairbrush and looks into a hand mirror.

#1982-33 *Super Teen Skipper Paper Doll.*

YEAR	NUMBER	PRICE	TITLE	MAKER	VALUE
1980	#1982-33	99¢	Super Teen Skipper Paper Doll	Whitman	$15.00

The same as #1980-3 except for the different stock number and price.

YEAR	NUMBER	PRICE	TITLE	MAKER	VALUE
1981	#1982-34	99¢	**Pretty Changes Barbie Paper Doll**	**Whitman**	**$20.00**

The Pretty Changes paper doll is very attractive with her short, blonde hair. The wigs and the original outfits are beautifully presented against colorful backgrounds.

Cover:

Pretty Changes Barbie #2598 in the original doll outfit with two small close-up pictures of Pretty Changes Barbie to her side. One of the small Barbie dolls has a long blonde wig and the other has a brunette wig.

Paper Doll:

One press-out doll: Pretty Changes Barbie #2598 with short, blonde hair wears a strapless, yellow one-piece swimsuit.

Outfits:

Four pages of press-out outfits.

#1982-42 Pretty Changes Barbie Paper Doll.

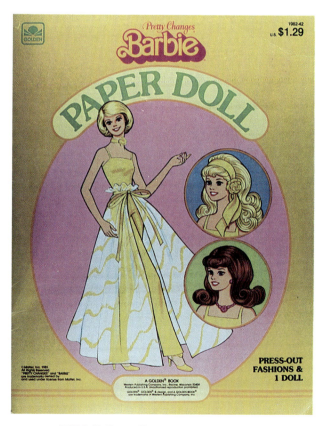

#1982-42 Pretty Changes Barbie Paper Doll (Note cover changes.)

Outfits (continued):
#2598 Pretty Changes doll outfit with skirt
#2598 Pretty Changes doll outfit pants

Beginner's Fashions: #1368 Wrap 'n Tie, #1370 Wrap, Snap, Tie!

Best Buy: #1352

Accessories:
Hats, hair extensions, and head scarf.

Cardboard Insert:
Punch out and assemble a tote to carry outfits. The paper doll is too large for the tote. Some clothes need to be folded to store in the tote.

Back Cover:
Three Pretty Changes Barbie dolls pose wearing a brunette wig, a short, blonde wig and a long, blonde wig.

YEAR	NUMBER	PRICE	TITLE	MAKER	VALUE
1981	#1982-42	$1.29	**Pretty Changes Barbie Paper Doll**	Whitman	$20.00

The same as #1982-34 except for the different stock number and price.

YEAR	NUMBER	PRICE	TITLE	MAKER	VALUE
1981	#1982-42	$1.29	**Pretty Changes Barbie Paper Doll**	Golden	$20.00

The same as #1982-34 except that Golden issued this book, and smaller Barbies appear on the cover.

YEAR	NUMBER	PRICE	TITLE	MAKER	VALUE
1982	#1982-43	$1.29	Western Barbie Paper Doll	Golden	$20.00

The attractive cover, unusual paper doll, and Dallas make this a clever paper doll book. The only negative is the lack of original Barbie doll outfits.

Cover:

Western Barbie #1757 wears her original Western Barbie doll outfit while riding her horse Dallas and waving with her hat.

Paper Doll:

Two press-out paper dolls: Western Barbie #1757 in her original Western doll outfit and Western Barbie #1757 in a blue and green one-piece swimsuit. Encircled in a lasso, the Western Barbie paper doll is sitting in an unusual position.

Outfits:

Four pages of press-out outfits. #1757 Western Barbie doll outfit

Accessories: Hats

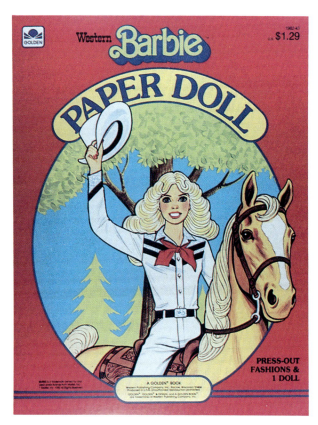

#1982-43 *Western Barbie Paper Doll.*

Western Barbie doll with paper doll (posed to ride her horse) from #1982-43 book.

Cardboard Insert:

Press-out Dallas and his saddle.

Back Cover:

Western Barbie and Western Ken dance arm-in- arm. The background holds an oval picture, a small Ken sitting on a barrel playing a fiddle. Two oval pictures in the background show Barbie and Ken riding horses and Ken riding a bronco.

YEAR	NUMBER	PRICE	TITLE	MAKER	VALUE
1982	#1983-43	$1.29	Golden Dream Barbie Paper Doll	Whitman	$15.00

Although the book has a limited number of original Barbie doll outfits, it beautifully illustrates the Golden Dream Barbie with gold fashions and wigs. The gold Western Barbie outfits and a Japanese kimono are charming.

Cover:

Golden Dream Barbie #1874 holds the edges of her full golden skirt. A smaller Golden Dream Barbie is in the background.

Paper Dolls:

One press-out doll: Golden Dream Barbie #1874 is in a one-piece strapless yellow and orange swimsuit.

Outfits:

Four pages of press-out outfits.
#1757 Western Barbie doll outfit (gold)
#1874 Golden Dream Barbie doll outfit
Designer Original: #1958 Golden Accent
    #1957 Gold Spun

Accessories:

Wigs, hairpieces, hat, flowers.

Cardboard Insert:

Press out and assemble a tote to carry the paper doll and outfits.

Back Cover:

Three small Golden Dream Barbie dolls model the Western Barbie doll outfit and two Golden Dream Barbie doll outfits.

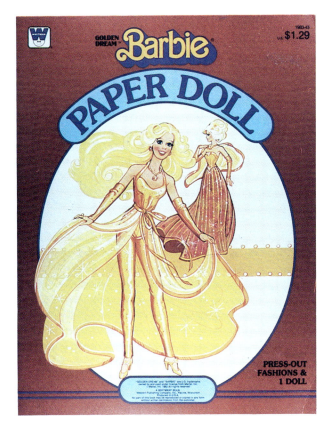

#1983-43 *Golden Dream Barbie Paper Doll*

YEAR	NUMBER	PRICE	TITLE	MAKER	VALUE
1983	#1982-44	$1.29	**Sunsational Malibu Barbie Paper Doll**	Golden	$15.00

An average paper doll book with a wide variety of sports fashions.

Cover:
Sunsational Malibu Ken #1088 and Sunsational Malibu Barbie #1067 at the beach. Ken holds a beach ball while Barbie applies sunscreen.

Paper Dolls:
Two press-out dolls: Sunsational Malibu Barbie #1067 wears the original doll swimsuit and Sunsational Malibu Ken #1088 wears red trunks and a white "Life Guard" shirt.

Outfits:
Four pages of press-out outfits.

Accessories: Hats.

Cardboard Insert:
Press out and assemble a beach umbrella.

Back Cover:
In a pool, Sunsational Malibu Ken splashes water toward Sunsational Malibu Barbie who rests on a towel beside him.

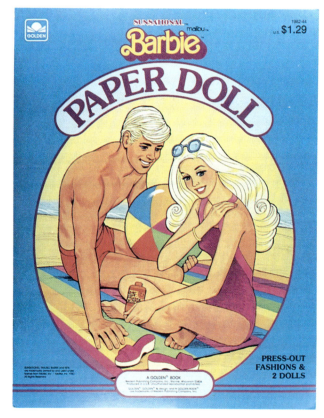

#1982-44 *Sunsational Malibu Barbie Paper Doll*

YEAR	NUMBER	PRICE	TITLE	MAKER	VALUE
1983	#1982-45	$1.29	**Angel Face Barbie Paper Doll**	Golden	$15.00

A good collection of My First Barbie outfits.

Cover:
Angel Face Barbie #5640 wearing the original doll outfit sits on a swing and looks at the sky.

Paper Doll:
One press-out paper doll: Angel Face Barbie #5640 wears a one-piece pink and white swimsuit with black trim.

Outfits:
Four pages of press-out outfits.
#5640 Angel Face original doll outfit

Fashion Classic Outfits:
#5700 Paint the Town Red
#5708 Ski Party Pink (deeper pink)

My First Barbie doll outfits:
#5609 My First Day of School (dark blue skirt)
#5610 My First Slumber Party (darker blue)
#5612 My First Ski Trip
#5613 My First Frills
#5614 My First Date

Fashion Fantasy:
#5536 Stripes for Stretching

Accessories: Hat and ball.

Cardboard Insert:
Punch out and assemble a tote to store and carry the doll and outfits.

Back Cover:
A beautiful Angel Face Barbie holds a pink rose.

#1982-45 *Angel Face Barbie Paper Doll*

YEAR	NUMBER	PRICE	TITLE	MAKER	VALUE
1983	#1982-46	$1.29	Twirly Curls Barbie Paper Doll	Golden	$15.00

Most fashions are not original Barbie doll outfits. However, the outfits are attractive because of the attached accessories such as a raincoat with an umbrella, a skirt and sweater outfit with books, several dresses with purses, and the babysitter's dress and apron with a girl toddler.

Cover:

Twirly Curls Barbie #5579 looks into a small compact mirror to improve her hairstyle. Twirly Curls Barbie is dressed in the original doll outfit.

Paper Doll:

One press-out paper doll: Twirly Curls Barbie #5579 wears a one-piece strapless pink swimsuit.

Outfits:

Four pages of press-out outfits. #5579 Twirly Curls doll outfit. #5315 Fashion Jean doll outfit (lighter blue jeans.)

Accessories: Hats.

Cardboard Insert:

Press out and assemble a tote to store and carry doll and outfits.

Back Cover:

Twirly Curls Barbie sits at her dressing table.

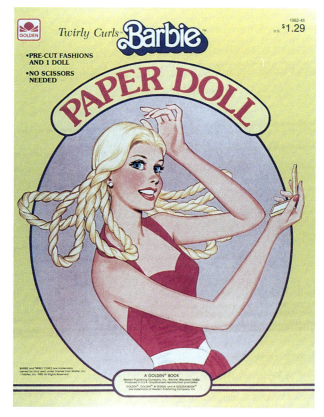

#1982-46 *Twirly Curls Barbie Paper Doll.*

YEAR	NUMBER	PRICE	TITLE	MAKER	VALUE
1983	#1836-43	$2.59	**Pink and Pretty Barbie Paper Doll Playbook**	Golden	$20.00

The paper doll and outfits are too small. The fashion show play scene is appropriate for Barbie as a model.

Cover:

Pink and Pretty Barbie #3554 wears sparkling jewelry with the original doll outfit. Three small square pictures at the bottom of the cover show items in the book; doll and clothes, a fashion show setting, and entrance play scene.

Paper Doll:

One punch-out paper doll: Pink and Pretty Barbie #3554 dressed in a strapless, one-piece pink swimsuit.

Outfits:

Four pages of punch-out outfits. #3554 Pink and Pretty doll outfit. #1757 Western doll outfit. #5315 Fashion Jeans doll outfit.

Accessories:

Hats, mirror, brush, make-up, and flowers.

Cardboard Inserts:

Punch-out and assemble a fashion show setting including a stage, piano, piano player, planters, and other accessories.

Inside Cover:

Punch out and assemble an entrance play scene which includes a stage curtain, audience, waiter, and two-piece orchestra.

Back Cover:

A doorman stands at the entrance to the fashion show theater.

#1836-43 *Pink and Pretty Barbie Paper Doll Playbook.*

YEAR	NUMBER	PRICE	TITLE	MAKER	VALUE
1983	#1983-44	$1.29	**Pink and Pretty Barbie Paper Doll**	Golden	$15.00

The Pink and Pretty outfit is accurately and beautifully illustrated. The paper doll and outfit pages have tiny pink roses in the background.

Cover:

A large colorful close-up of Pink and Pretty Barbie #3554 wearing the original doll outfit.

Paper Doll:

One press-out paper doll: Pink and Pretty Barbie #3554 dressed in a one-piece strapless pink swimsuit.

Outfits:

Four pages of press-out outfits: two #3554 Pink and Pretty doll outfits.

Accessories: None.

Cardboard Insert:

Press out and assemble a tote to store and carry doll and outfits.

Back Cover:

Two Pink and Pretty Barbie dolls stand beside each other dressed in different pieces of the original outfit.

#1983-44 *Pink and Pretty Barbie Paper Doll*

YEAR	NUMBER	PRICE	TITLE	MAKER	VALUE
1984	**#1527**	**$1.29**	**Barbie and Ken Paper Doll**	**Golden**	**$20.00**

This book contains only original Barbie and Ken doll outfits. Barbie and Ken have four sets of matching formal wear. The Angel Face, Crystal, and Twirly Curls Barbie doll gowns and two Collector Series gowns are beautifully and accurately illustrated. An excellent representation of the '80s glamorous fashions and a wonderful collector's book.

Cover:

Crystal Barbie #4859 dances with Crystal Ken #4895 against a green background. Barbie is dressed in Collector Series III Silver Sensation #7438 and Ken wears the original doll outfit with a pink bow tie, boutonniere, and cummerbund to match Barbie doll's gown.

Paper Doll:

Two press-out paper dolls: Crystal Barbie #4859 wears a one-piece pink swimsuit from Dance Sensation #7218 and Sun Gold Malibu Ken #1088 wears blue trunks and red T-shirt.

Outfits: Four pages of press-out outfits.

Barbie

Twice as Nice Reversible Fashions

#4821 Double Date
#4859 Crystal Barbie original doll gown
#5579 Twirly Curls original doll gown
#5640 Angel Face Barbie original doll outfit
#7092 Collector Series II Springtime Magic Fashion

Spectacular Fashions

#7218 Dance Sensation short dress
#7438 Collector Series III Silver Sensation Fashion.

Ken

#4077 Dream Date Ken original doll suit
#4077 Dream Date Ken original doll suit (white coat and pink bow tie)
#4898 Crystal Ken original doll suit (blue suit, orchid tie and boutonniere)
Designer Collection: #5651 Date Night (red cummerbund and boutonniere)

Accessories: Hat.

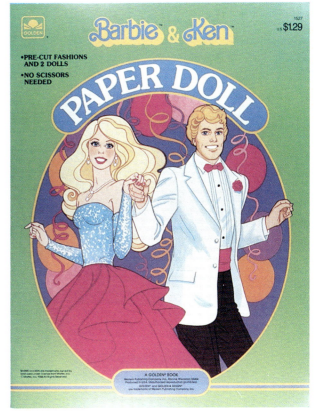

#1527 Barbie and Ken Paper Doll

Back Cover:

Barbie and Ken dance to rock and roll music with Barbie wearing #4821 Double Date and blond Ken wearing #4077 Dream Date doll outfit with a white coat, red tie, and cummerbund.

Collector Series III, Silver Sensation outfit with gown shown on the cover of #1527 book.

YEAR	NUMBER	PRICE	TITLE	MAKER	VALUE
1984	#1985-51	$1.29	**Barbie and Ken Paper Doll**	Golden	$20.00

The same as #1527 except for the darker green background cover and a different stock number.

YEAR	NUMBER	PRICE	TITLE	MAKER	VALUE
1984	#1731	$1.29	**Barbie Christmas Time Paper Doll**	Golden	$20.00

Candy canes, wreaths, bells, and holly appear on all outfit pages. Many outfits include accessories such as Christmas gifts, decorations, Christmas carol book, cookies, poinsettia, and socks. A very special paper doll book because of the Christmas theme and the pretty paper doll and outfits.

Cover:

Happy Birthday Barbie #5614 places an ornament on a Christmas tree with mistletoe above her head. Barbie is dressed in My First Date #1922.

Paper Doll:

One press-out paper doll: Happy Birthday Barbie #5614 wears a red and white teddy trimmed in red.

Outfits:

Four pages of press-out outfits.

Collector Series I: #4277 Heavenly Holidays

Fashion Fantasy: #4811 Curtain's Up

Twice as Nice Reversible Fashions
#4821 Double Date
#4824 Double Dazzle

My First Barbie Fashions
#5610 My First Slumber Party
#5614 My First Date

Accessories: None.

Cardboard Insert:

Punch out and decorate a Christmas tree with ornaments by placing their tabs into slits in the tree. The ornaments include a rocking horse, doll, gingerbread man, angel, snowman, bird, Santa, bell, and two tin soldiers.

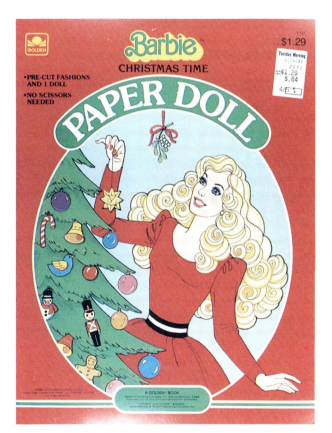

#1731 *Barbie Christmas Time Paper Doll.*

Back Cover:

Large close-up of Barbie holding her Christmas list and many packages.

YEAR	NUMBER	PRICE	TITLE	MAKER	VALUE
1984	#1982-47	$1.29	**Barbie Fantasy Paper Doll**	Golden	$15.00

A wonderful book with original doll fashions from 1983.

Cover:

Pink and Pretty Barbie #3554 wears Collector Series I, Heavenly Holidays #4277 and sits with an open book, pad, and pencil.

Paper Doll:

One press-out doll: Pink and Pretty Barbie #3554 dressed in pink undergarments.

Outfits:

Four pages of press-out outfits.

Collector Series I: #4277 Heavenly Holidays

Fashion Fun
#5711 Summer Party
#5712 Romantic Nights
#5713 Lunch Date
#5717 City Fun
#5719 Backyard Barbecue
#5543 Summertime Fun
#5720 Sophisticated Lady
#5898 Candlelight Nights

Fantasy Fashion
#5539 Stripes for Stretching
#5542 City Sailor
#5544 Party in Purple
#5545 Summer Sparkle
#5546 Shine at the Party
#5547 Dream Time
#5548 Evening Rose

Accessories: Hat.

Cardboard Insert:

Press out and assemble a tote to carry doll and outfits.

Back Cover:

Three circles picture three Barbies dressed in #5542 City Sailor, #5713 Lunch Date, and #5711 Summer Party.

#1982-47 *Barbie Fantasy Paper Doll* wearing Heavenly Holidays outfit on cover.

Collector Series I, Heavenly Holiday outfit with paper doll outfit from #1731 book.

YEAR	NUMBER	PRICE	TITLE	MAKER	VALUE
1984	#1983-46	$1.29	**Crystal Barbie Paper Doll**	Golden	$15.00

An excellent book picturing Crystal Barbie paper doll and the original outfits from 1984.

Cover:

Crystal Barbie #4598 snaps her fingers wearing the original doll outfit.

Paper Doll:

One press-out doll: Crystal Barbie #4598 wears a white, one-piece swimsuit with a pink waistband in a fabric similar to Crystal Barbie doll's gown.

Outfits:

Four pages of press-out outfits.

Outfits:

Collector Series II: Springtime Magic #7092

Fashion Fantasy
#4810 A Little Luxury
#4815 Letter Perfect
#4817 Jazzdancin' (blue)

Fashion Fun
#4801 Fun in the Sun (red and blue)
#4802 Golden Mine
#4804 Everyday Outing
#4807 Soft and Lacy

Twice As Nice: #4824 Double Dazzle

My First Barbie Fashions
#4868 My First Party Dress
#4870 My First Day at the Park

Spectacular Fashions
#7218 Dance Sensation

Accessories: Hat.

Cardboard Insert:
Press out and assemble a tote to carry doll and outfits.

Back Cover:
A large close-up of a winking Crystal Barbie.

#1983-46 *Crystal Barbie Paper Doll.*

Collector Series II, Springtime Magic outfit with paper doll outfit from #1983-46 book.

YEAR	NUMBER	PRICE	TITLE	MAKER	VALUE
1985	#1522	$1.29	**Great Shape Barbie Paper Doll**	Golden	$15.00

An especially fine book with fitness and exercise as the theme. Most outfits are original Barbie, Ken, and Skipper doll outfits.

Cover:

Great Shape Barbie #7025, Great Shape Ken #7318, and Great Shape Skipper #7417 work out in their original doll outfits.

Paper Dolls:

Three press-out dolls: Great Shape Barbie #7025 wears a one-piece pink and purple swimsuit, Great Shape Skipper #7417 wears Pool Party #4876 and Great Shape Ken #7318 wears blue trunks and shirt from original Great Shape outfit.

Outfits:

Four pages of press-out outfits.

Barbie

#7025 Great Shape doll outfit

Fashion Fantasy: #4817 Jazzdancin'

My First Barbie Fashions
    #4872 My First Picnic

Twice as Nice: #4825 In & Outfitted

Skipper

#7417 Great Shape doll outfit
#4876 Pool Party
#4880 Sightseeing

Ken

#7318 Great Shape doll outfit
Twice As Nice
    #4885 Ship-Shape
    #4886 Double Play

Accessories: None.

Back cover:

Great Shape Barbie, Ken, and Skipper doing various forms of exercise.

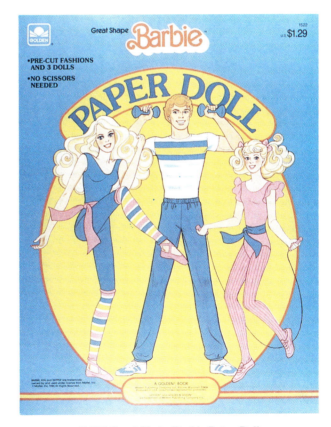

#1522 *Great Shape Barbie Paper Doll.*

YEAR	NUMBER	PRICE	TITLE	MAKER	VALUE
1985	#1982-49	$1.29	Great Shape Barbie Paper Doll	Golden	$15.00

The same as #1522 except for the different stock number.

YEAR	NUMBER	PRICE	TITLE	MAKER	VALUE
1985	#1525	$1.29	Peaches 'n Cream Barbie Paper Doll	Golden	$15.00

The colorful cover and glamorous gowns represent the Peaches 'n Cream Barbie doll accurately.

#1938-48 or #1525 *Peaches 'n Cream Barbie Paper Doll.*

Cover:

With a green background, smiling Peaches 'n Cream Barbie #7926 sits near a tree holding a bluebird.

Paper Doll:

One press-out doll: Peaches 'n Cream Barbie #7926 dressed in a strapless pink swimsuit with white heart pattern and white lace trim.

Outfits:

Four pages of press-out outfits.
#7926 Peaches 'n Cream Barbie original doll outfit.

Spectacular Fashions: #9144

Active Fashions: #7912, #7913, #7917

Accessories: Hats and purse.

Cardboard Insert:

Press out and assemble a tote to carry doll and outfits.

Back Cover:

One Peaches 'n Cream Barbie dressed in Active Fashion #7917 lounges on several oversized pillows. Another Peaches 'n Cream Barbie dressed in Spectacular Fashions #9146 with a long fur stole holds the leash of a white poodle "sitting up" on a pink pillow.

YEAR	NUMBER	PRICE	TITLE	MAKER	VALUE
1985	#1983-48	$1.29	Peaches 'n Cream Barbie Paper Doll	Golden	$15.00

The same as #1525 except for a dark green cover and the different stock number.

YEAR	NUMBER	PRICE	TITLE	MAKER	VALUE
1985	#1521	$1.29	Day-to-Night Barbie Paper Doll	Golden	$20.00

A wonderful Day-to-Night Barbie book with all the Day-to-Night Fashions for work and play.

Cover:

Day-to-Night Barbie #7929 wears the original doll outfit and poses in front of the city sky-scrapers.

Paper Dolls:

One press-out doll: Day-to-Night Barbie #7929 dressed in a one-piece pink swimsuit

Outfits:

Four pages of press-out outfits.
#7929 Day-to-Night Fashions doll outfits suit and dress
#9081 Dress Designer (white jacket, two fashions)
#9082 Dancer (two fashions)
#9083 Business Executive (two fashions)
#9084 News Reporter (two fashions)
#9085 Teacher (two dresses)

Accessories: Hats.

Cardboard Insert:

Press out and assemble a tote to store doll and outfits.

Back Cover:

A close-up of Day-to-Night Barbie holding a pad
and pencil. Two other Day-to-Night Barbies stand to her side. One Barbie who wears the Day-to-Night suit carries her attaché case and waves while the other Barbie wears the Day-to-Night evening dress.

#1982-48 or #1521 *Day-to-Night Barbie Paper Doll.*

YEAR	NUMBER	PRICE	TITLE	MAKER	VALUE
1985	#1982-48	$1.29	Day-to-Night Barbie Paper Doll	Golden	$20.00

The same as #1521 except for the different stock number.

YEAR	NUMBER	PRICE	TITLE	MAKER	VALUE
1986	#1528	$1.29	Barbie and the Rockers Paper Doll	Golden	$15.00

All the original Rocker outfits are accurately pictured with the colorful outfits, make-up and full hair. The paper dolls' poses make them look active.

Cover:

Barbie and the Rockers, Barbie #1140, Dee Dee #1141, Dana #1196, Diva #2427, and Derek #2428, singing and rockin' dressed in the original doll outfits.

Paper Dolls:

Five press-out dolls: Each doll wears a one-piece swimsuit and a hair decoration. Barbie #1140 has a pink swimsuit with a large purple bow, Diva #2427 has a pink swimsuit with a headband, Dee Dee #1141 wears a gold swimsuit with large orange bow, Dana #1196 wears a red swimsuit with large yellow bow, and Derek #2428 wears blue trunks and yellow top.

Outfits:

Four pages of press-out outfits.

Rocker Outfits: #1165, #1167, #1176, #1177, #2688, #2689, #2691

Rocker Barbie #1140 original doll outfit
Rocker Dana #1196 original doll outfit
Rocker Derek #2428 original doll outfit
Rocker Diva #2427 original doll outfit
Rocker Dee Dee #1141 original doll outfit

Accessories: None.

Back cover:

Rocker Barbie #1140 sings into her microphone while Rocker Derek #2428 plays the guitar.

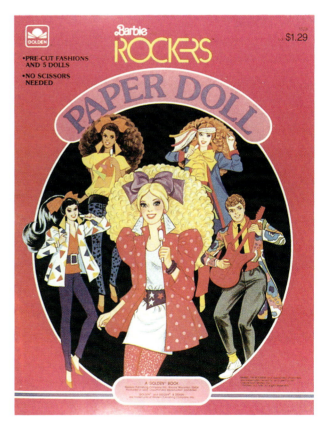

#1528 Barbie and the Rockers Paper Dolls

YEAR	NUMBER	PRICE	TITLE	MAKER	VALUE
1986	#1528-1		Barbie and the Rockers Paper Doll	Golden	$15.00

The same as #1528 except a different stock number and without a price.

YEAR	NUMBER	PRICE	TITLE	MAKER	VALUE
1986	#1523	$1.29	**Tropical Barbie Paper Doll**	Golden	$15.00

This book has a pretty cover and colorful paper dolls.

Cover:

Tropical Barbie #1017 holds a white Tahiti Bird #2064. Tropical Miko #2056, Tropical Ken #1020, and Tropical Skipper #1021 sit in the background. All Tropical dolls wear their original outfits.

Paper Dolls:

Four press-out dolls: Blonde Tropical Barbie #1017, Tropical Ken #1020, Tropical Miko #2056 and Tropical Skipper #1021 are all dressed in their original swimwear. Barbie, Miko, and Skipper have very long hair.

Outfits:

Four pages of press-out outfits.
#1017 Tropical Barbie original swimsuit
#2056 Tropical Miko original swimsuit
#1021 Tropical Skipper original swimsuit

Accessories:

Hats, flowers, and lei.

Back Cover:

Press-out Tahiti Bird #2064 fits into slit on perch.

#1523 *Tropical Barbie Paper Doll.*

YEAR	NUMBER	PRICE	TITLE	MAKER	VALUE
1986	#1523-1		**Tropical Barbie Paper Doll**	Golden	$15.00

The same as #1523 except the different stock number and without a price.

YEAR	NUMBER	PRICE	TITLE	MAKER	VALUE
1987	#1537	$1.29	Jewel Secrets Barbie Paper Doll	Golden	$15.00

All paper doll fashions are Jewel Secrets outfits including different versions of the same fashions.

Cover:

The four Jewel Secrets dolls, Barbie #1737, Ken #1719, Whitney #3179, and Skipper #3133, pose for a group photo in their original doll clothes.

Paper Dolls:

Four press-out dolls: All three Jewel Secrets, Barbie #1737, Whitney #3179, and Skipper #3133, wear one-piece strapless swimsuits in pink, blue, and orchid, respectively. Jewel Secrets Ken #1719 wears blue trunks and a white top. All dolls look thin compared to dolls pictured in other paper doll books.

Outfits:

Four pages of press-out outfits.

#1859 Jewel Secrets Fashion
#1860 Jewel Secret Fashion
#1737 Jewel Secrets Barbie original doll outfit

#3133 Jewel Secrets Skipper original doll outfit
#1719 Jewel Secrets Ken original doll outfit
#3179 Jewel Secrets Whitney original doll outfit

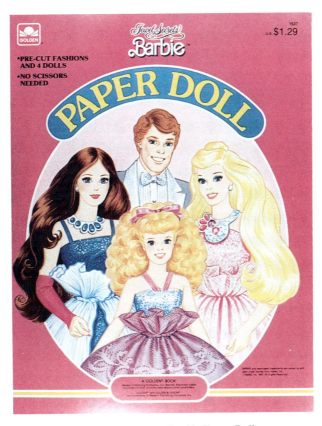

#1537 *Jewel Secrets Barbie Paper Doll.*

Outfits (continued):
   Jewel Secrets Fashions: #1859, #1860, #1861, #1862, #1865

Accessories: None

Back Cover:
There is a press-out "jewel" choker for little girl to wear.

Jewel Secrets Barbie doll with #1537 book.

YEAR	NUMBER	PRICE	TITLE	MAKER	VALUE
1987	#1537-1		Jewel Secrets Barbie Paper Doll	Golden	$15.00

The same as #1537 except the different stock number and without a price.

YEAR	NUMBER	PRICE	TITLE	MAKER	VALUE
1988	**#1500**		**Perfume Pretty Barbie Paper Doll**	**Golden**	**$20.00**

An outstanding book with all the elegant Perfume Pretty outfits. Ken doll's tuxedos include small gifts while Barbie doll's gowns include tiny bouquets. One paper doll page has two press-out bracelets for a little girl to wear. A beautiful representation of the Perfume Pretty dolls.

Cover:

Perfume Giving Ken #4554 presents Perfume Pretty Barbie #4551 a bottle of perfume. In the background, Perfume Pretty Whitney #4557 holds a small bouquet.

Paper Dolls:

Three press-out dolls: Perfume Pretty Barbie #4551 wears a one-piece pink swimsuit. Muscular Perfume Giving Ken #4554 wears blue trunks and burgundy T-shirt. Perfume Pretty Whitney #4557 wears a patterned one-piece swimsuit.

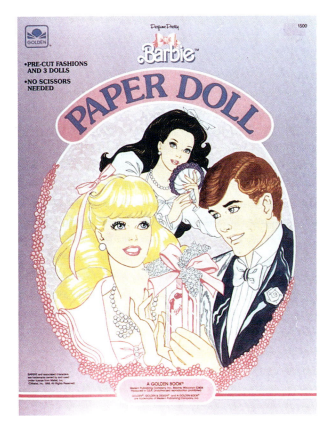

#1500 *Perfume Pretty Barbie Paper Doll.*

Outfits:

Four pages of press-out outfits.
#4551 Perfume Pretty Barbie doll fashions (two)
#4557 Perfume Pretty Whitney doll fashion
#4554 Perfume Giving Ken doll outfit
#4554 Perfume Giving Ken doll outfit
    (blue and white)

Perfume Pretty Fashions

#4620, #4621, #4622, #4623, #4624 (pink), #4625 (white)

Private Collection

Ken: #4508 (pink bowtie and cummerbund)

Accessories: None.

Back Cover:

There is a press-out "pearl" choker for little girl to wear.

YEAR	NUMBER	PRICE	TITLE	MAKER	VALUE
1989	#1537-2		SuperStar Barbie Paper Doll	Golden	$20.00

A beautiful paper doll book with all the glamourous SuperStar gowns.

Cover:

A close-up of SuperStar Barbie #1604 holding an award statuette. SuperStar Barbie wears the original doll outfit with large, dangling earrings with stars.

Paper Doll:

One press-out paper doll: SuperStar Barbie #1604, with her hair blowing to one side, wears a one-piece strapless blue swimsuit with white stars.

Outfits:

Four pages of press-out outfits.
#1604 Super Star Barbie original doll outfit

Super Star Fashion Assortment
#3300 (two), #3301, #3303 (two), #3304, #3305, #3306

Paris Pretty Fashions: #1908

Accessories:

Hat and hairpiece.

Cardboard Insert:

Press out and assemble a tote to carry doll and outfits.

Back Cover:

Press out and assemble a stage for SuperStar Barbie.

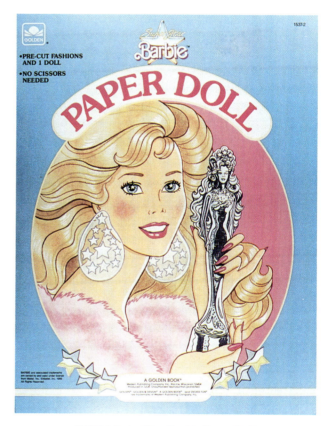

#1537-2 *SuperStar Barbie Paper Doll.*

Paper Doll Books of the 1990s

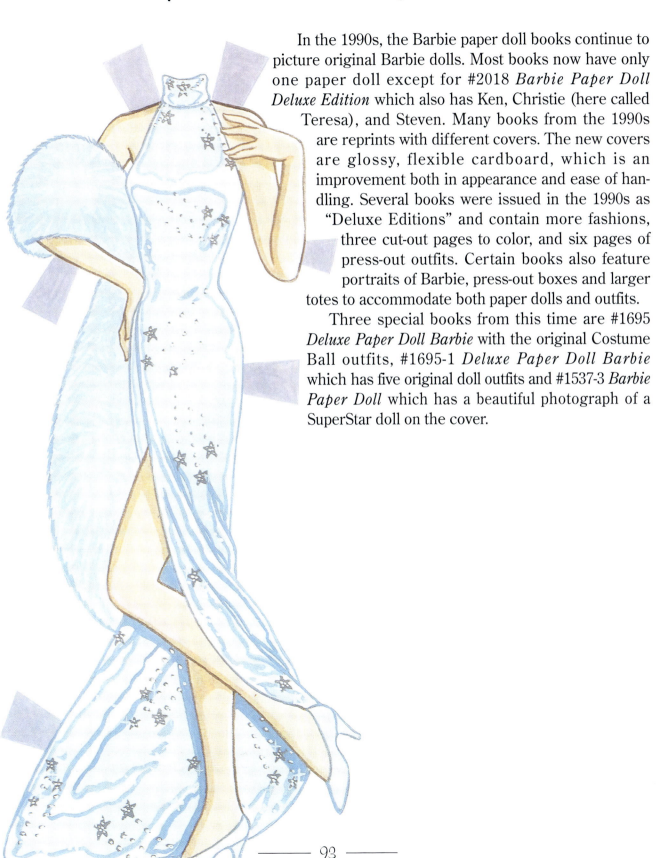

In the 1990s, the Barbie paper doll books continue to picture original Barbie dolls. Most books now have only one paper doll except for #2018 *Barbie Paper Doll Deluxe Edition* which also has Ken, Christie (here called Teresa), and Steven. Many books from the 1990s are reprints with different covers. The new covers are glossy, flexible cardboard, which is an improvement both in appearance and ease of handling. Several books were issued in the 1990s as "Deluxe Editions" and contain more fashions, three cut-out pages to color, and six pages of press-out outfits. Certain books also feature portraits of Barbie, press-out boxes and larger totes to accommodate both paper dolls and outfits.

Three special books from this time are #1695 *Deluxe Paper Doll Barbie* with the original Costume Ball outfits, #1695-1 *Deluxe Paper Doll Barbie* which has five original doll outfits and #1537-3 *Barbie Paper Doll* which has a beautiful photograph of a SuperStar doll on the cover.

YEAR	NUMBER	PRICE	TITLE	MAKER	VALUE
1990	**#1502**		**Barbie Paper Doll**	**Golden**	**$15.00**

The outfits range from the glamorous Dance Magic gowns to the Wet 'n Wild fashions. The outfit pages are decorated with flowers while the paper doll page has large colorful cacti.

Cover:

Western Fun Barbie #9932 dressed in the original doll outfit stands among a variety of desert cacti.

Paper Doll:

One press-out doll: Western Fun Barbie #9932 wears a one-piece strapless swimsuit in a multi-color print.

Outfits:

Four pages of press-out outfits.
#2249 Home Pretty Barbie doll original outfit
#4836 Dance Magic Barbie doll original outfit
#1054 Wet 'n Wild Fashions
Barbie Cool Mix: #2132 – #2146
#2496 All Stars Fashion Assortment
#7392 Dance Magic Fashion Assortment
#9161 Barbie Dinner Date Fashion
#9253 My First Barbie Fashions
#9951 Western Fun Fashions
#9964 Fun to Dress
#9971 Barbie Fashion Finds

Accessories:

Hat and flowers for the hair

Cardboard Insert:

Press out and assemble a tote to carry doll and outfits.

Back Cover:

A portrait of Western Fun Barbie #7392 with a punch-out picture frame and stand.

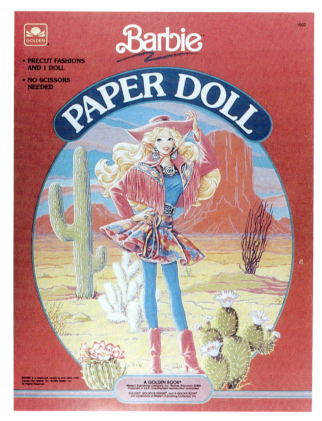

#1502 *Barbie Paper Doll* (Western).

YEAR	NUMBER	PRICE	TITLE	MAKER	VALUE
1990	#1502-1		**Barbie Paper Doll**	**Golden**	**$15.00**

An excellent book because of the beautiful wedding gown and brightly colored original outfits. The concept of coloring some outfits is introduced.

Cover:
Wet 'n Wild Barbie #4103 dressed in the original doll outfit points to three small Barbie dolls dressed in Wedding Fantasy #2125, Ice Capades outfit #4083 and Western Fun outfit #9952.

Paper Doll:
One press-out doll: Wet 'n Wild Barbie #4103 wears the original doll swimsuit.

Outfits:
Four pages of press-out outfits.
#2125 Wedding Fantasy Barbie original doll outfit
#2751 Barbie and the Beat original doll outfit
#9942 My First Barbie doll original outfit
All Star Fashion Assortment: #2609 – #2553
#4083 Ice Capades Barbie Fashions
#4573 Barbie and the Beat
#7394 Dance Magic Fashion

Western Fun Fashions: #9952 – #9953.

Accessories:
Cowgirl hat, My First Barbie crown and bridal veil.

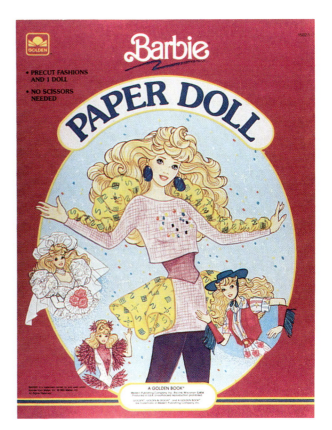

#1502-1 *Barbie Paper Doll* (Wet 'N Wild).

Cardboard Insert:
Press out and assemble a tote to carry doll and outfits.

Back Cover:
Three punch-out clothes to color. Decorate the clothes with crayons, markers or paints or add buttons, stickers, lace or ribbon for trim.

YEAR	NUMBER	PRICE	TITLE	MAKER	VALUE
1990	#1523-2		**Barbie Paper Doll**	Golden	$15.00

The book features a diversity of outfits from the 1990 Barbie fashion assortment. The paper doll and the colorful fashions make this an exceptional paper doll book.

Cover:

Beach Blast Barbie #3237 holds a red rose that matches the red rose on her large yellow hat from the outfit she wears, Paris Pretty Fashion #1909.

Paper Doll:

One press-out doll: Beach Blast #3237 in her original doll outfit.

Outfits:

Four pages of press-out outfits.

#1953 Garden Party Barbie original doll outfit
#1304 Dinner Date Fashion
#1464 Style Magic Fashion Assortment
#1532 Barbie Weekend Collection
#1595 Animal Lovin' Fashion Assortment

Paris Pretty Fashions: #1907 – #1909
#3322 Cool Times Fashion Assortment
Dance Club Fashions: #3567 – #3566
#4125 Barbie Pretty Choices
#4334 The Jean Look

#1523-2 *Barbie Paper Doll* (Beach Blast).

Fashions from the #1523-2 book. (Note cover outfit.)

Accessories:

Hats and hair bowties.

Cardboard Insert:

Press out and assemble a tote to carry the doll and outfits.

Back Cover:

Press out a box to store trinkets or jewelry.

Left: #1909 Paris Pretty Fashion outfit from Pret-a-Porter outfit box. Right: Close-up of matching paper doll outfit from #1523-2 *Barbie Paper Doll* book.

YEAR	NUMBER	PRICE	TITLE	MAKER	VALUE
1990	#1690		**Deluxe Paper Doll Barbie**	Golden	$15.00

The cover has an unflattering picture of the beautiful Dance Magic doll. The press-out pages include many beautiful red and pink outfits on pages with pink stars. The book has an increase in outfit pages with the three additional pages to color.

Cover:

The elegant Dance Magic Barbie #4836 wears the original doll outfit.

Paper Doll:

One press-out doll: Dance Magic #4836 dressed in All Star Sports Club original outfit #9099.

Outfits:

Three pages of cut-out outfits to color and six pages of press-out outfits. One cut-out storage envelope for doll and clothes.
#4836 Dance Magic original doll outfit
#9932 Western Fun original doll outfit
#4961 Private Collection Fashion Assortment
#7391 Dance Magic Fashion Assortment
Western Fun Fashion: #9950 – #9954
My First Barbie Fashion Assortment:
    #9245 – #9247

To Color:

Private Collection #4957 – #4959
#7394 Dance Magic Fashion Assortment
#9265 My First Barbie Fashion Assortment

Accessories:

Hats and a hair ornament.

#1690 *Barbie Deluxe Paper Doll* (Dance Magic).

YEAR	NUMBER	PRICE	TITLE	MAKER	VALUE
1990, 1991	**#1502-2**		**Barbie Paper Doll**	**Golden**	**$15.00**

The same as #1502 from 1990 except for a different cover.

#1502-2 *Barbie Deluxe Paper Doll* (Home Pretty).

Cover:

A photograph of Home Pretty Barbie #2249 dressed in the original doll outfit.

Back Cover:

A portrait of Western Fun Barbie #7392 with a punch-out picture frame and stand.

YEAR	NUMBER	PRICE	TITLE	MAKER	VALUE
1991	**#1695**		**Deluxe Paper Doll Barbie**	**Golden**	**$15.00**

Although many outfits are not original doll outfits, the beautiful cover and the collection of Costume Ball Fashions make the book special.

Cover:

A photograph of Dance Magic Barbie #4836 at night in a city dressed in Private Collection Fashion #7113.

Paper Doll:

One press-out doll: A photograph of Wet 'n Wild Barbie #4136 dressed in Wet 'n Wild Teresa doll swimsuit.

Outfits:

Three pages of cut-out outfits to color and six pages of press-out outfits. One cut-out storage envelope for the doll and clothes.

Private Collection Fashion Assortment:
#7092, #7100, #7113

Costume Ball Fashion Assortment:
#7763 Mermaid
#7764 Genie
#7765 Bunny
#7766 Bird

To Color:
Flower Surprise Fashion Assortment.

Accessories:
Hat and hairpiece.

#1695 *Barbie Deluxe Paper Doll* (Dance Magic).

YEAR	NUMBER	PRICE	TITLE	MAKER	VALUE
1991	#1695-1		**Deluxe Paper Doll Barbie More Fashions**	Golden	$15.00

A beautiful book because of the Costume Ball Barbie photograph cover and the five original doll outfits. The book has several pink outfits (Barbie doll's favorite color) with pink hearts decorating the outfit pages.

Cover:

A large photograph of Costume Ball Barbie #7123 dressed in the original doll outfit.

Paper Doll:

One press-out doll: Fashion Play Barbie #9629 dressed in original doll swimsuit.

Outfits:

Three pages of cut-out outfits to color and six pages of press-out outfits. One cut-out storage envelope for the doll and clothes.

Outfits:
#4833 – #4837 My First Barbie Fashions
#4744 Barbie Fashion Wraps Fashions
#4937 Barbie Dinner Date
#5931 – #5934: Barbie Flower Surprise Fashion
    Assortment
#5940 Hawaiian Fun Barbie doll original outfit
#7123 Costume Ball Barbie doll original outfit
#7913 Happy Birthday Barbie doll original outfit
#9423 All American Barbie doll original outfit
#9443 All American Barbie Fashion Assortment (2)
#9725 Lights and Lace Barbie doll original outfit

To Color:
#2483 My First Barbie Gift Set Outfit
#4853 – #4860: My First Barbie Fashions
#4936 Dinner Date Fashions
#4938 Dinner Date Fashions
#7765 Costume Ball Fashion Assortment
#9608 Wedding Day Midge, Barbie doll
    original outfit

Accessories:
Hat and hairpiece.

#1695-1 *Barbie Deluxe Paper Doll* (Costume Ball).

YEAR	NUMBER	PRICE	TITLE	MAKER	VALUE
1991, 1989	**#1537-3**		**Barbie Paper Doll**	**Golden**	**$15.00**
		The same as #1537-2 except for a different cover and different tote.			

Cover:
A close-up photograph of SuperStar Barbie #1604 dressed in the original doll outfit with large, dangling clear plastic earrings with stars.

#1537-3 *Barbie Paper Doll* (SuperStar).

Backcover of #1537-3 showing Barbie doll's press-out stage.

YEAR	NUMBER	PRICE	TITLE	MAKER	VALUE
1991, 1990	#1690-1		**Deluxe Paper Doll Barbie**	**Golden**	**$15.00**

The same as #1690 except for a different cover.

Cover:

Dance Magic Barbie #4836 wearing the original doll gown in subdued lighting.

Back Cover:

Press-out Dance Magic #4836 dressed in All Star Sports Club outfit #9099.

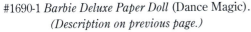

#1690-1 *Barbie Deluxe Paper Doll* (Dance Magic).
(Description on previous page.)

#1502-3 *Barbie Paper Doll* (All American Barbie).

YEAR	NUMBER	PRICE	TITLE	MAKER	VALUE
1992	#1502-3		**Barbie Paper Doll**	Golden	$15.00

This book has three, beautiful, original fashions and other lovely, pastel fashions similar in style to the original outfits of the time. A pretty cover and paper doll; however, the SuperStar position of the arms does not flatter the outfits. Pretty flowers are scattered around the outfit pages.

Cover:
A pretty close-up photograph of All American Barbie #9423 dressed in the original doll outfit.

Paper Doll:
One press-out doll: A photograph of a doll with SuperStar arms which is not a regular issue Barbie because of the change in hair style. Barbie wears Fancy Frills lingerie #2976.

Outfits:
Four pages of press-out outfits.
#2717 Private Collection   #4679 Sparkle Eyes Barbie Fashion Assortment
#2969 Great Date Fashion (with wig)

Accessories:

Bows and a hairpiece.

Cardboard Insert:

Press out and assemble a tote to carry the doll and outfits.

Back Cover:

Press out a box for jewelry storage.

YEAR	NUMBER	PRICE	TITLE	MAKER	VALUE
1992	#1690-2		Deluxe Paper Doll Barbie	Golden	Available

The cover and paper doll are pretty but the SuperStar position of the arms does not flatter the outfits. Pretty pink hearts and stars are scattered decorations on the outfit pages.

#1690-2 *Barbie Deluxe Paper Doll* (Sparkle Surprise).

Cover:

A photograph of Sparkle Surprise Barbie #3149 dressed in the original doll outfit.

Paper Doll:

One press-out doll: A photograph of Barbie doll which is not any regular issue Barbie. Barbie wears Fancy Frills lingerie #2976.

Outfits:

Six pages of press-out outfits, and three pages of cut-out and color outfits. Cut-out storage envelope for the doll and clothes.

#3839 My First Barbie Doll original doll outfit
#2482 Sparkle Eyes Barbie Doll original doll outfit (blue)
#2970 Great Date Fashion
#2977 Fancy Frills Lingerie Assortment

To Color:

#2972 Great Date Fashions

Accessories:

Wigs, hairpieces, hairbows and hats.

YEAR	NUMBER	PRICE	TITLE	MAKER	VALUE
1993	#1502-4		Barbie Paper Doll	Golden	$15.00

While the Christmas theme and Holiday Barbie are beautiful, the colors and styles of the unknown, unusual outfits in a myriad of bright colors are different from the typical glamorous Barbie fashions. The cover of the book does not represent the inside.

Cover:

J.C. Penney's exclusive 1993 Happy Holiday Barbie stands at the foot of a staircase decorated for Christmas. Barbie is dressed in the original doll outfit.

Paper Doll:

One press-out doll: Fun to Dress Barbie #3240 in two-piece orchid lingerie.

Outfits:

Four pages of press-out outfits.

Accessories:

Hats and bows.

Cardboard Insert:

Press out and assemble a tote to carry the doll and outfits.

Back Cover:

Press out and assemble a box to store trinkets or jewelry.

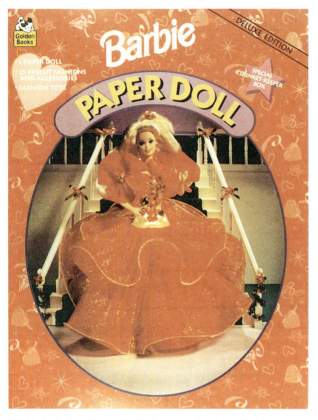 

#1502-4 *Barbie Paper Doll* features the 1993 J.C. Penney's Happy Holiday Barbie Doll on the cover. *(Description on previous page.)*

#1502-5 *Barbie Paper Doll*
(Romantic Bride)

YEAR	NUMBER	PRICE	TITLE	MAKER	VALUE
1993, 1990	#1502-5		**Barbie Paper Doll**	Golden	$15.00
The same as #1502-1 except for the different cover.					

Cover:

A beautiful photograph of Romantic Bride Barbie #1861 dressed in the original doll outfit.

YEAR	NUMBER	PRICE	TITLE	MAKER	VALUE
1994	#2018		Barbie Paper Doll Deluxe Edition	Golden	Available

The book has an interesting cover with attractive paper dolls and a complete collection of the Great Date Fashion Assortment. The Teresa paper doll looks like Christie.

Cover:

A photograph of four dolls walking past a cafe include Troll Barbie #10257, Sparkle Surprise Ken #3149, Sun Sensation Christie #1394 and Sun Sensation Steven #1396. The dolls are dressed in #2162 Magic Talk Barbie Fashion, #2090 Magic Talk Club, and #2188 Teen Talk Casual Fashion Assortment, respectively.

Paper Dolls:

Four press-out paper dolls: The paper dolls are photographs of the Glitter Beach dolls — Barbie #3602, Christie #4907, Ken #4904, and Steven #4918. Barbie wears Fancy Frills Lingerie Assortment #2978, Christie wears a one-piece flowered strapless swimsuit, and Ken and Steven wear swimming trunks.

#2018 *Barbie Paper Doll Deluxe Edition.*

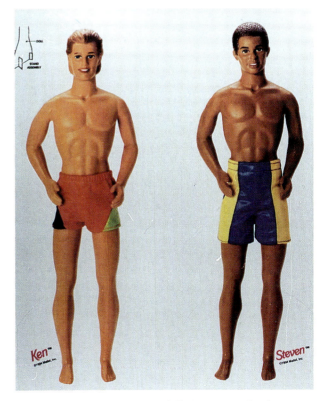

Ken and Steven paper dolls from #2018 book.

Outfits:
#10963 Locket Surprise Barbie doll outfit
Magic Talk Friends Fashion Assortment
#2162
#2193 Barbie – #2234 Ken matching outfits
Ken and Barbie Great Date Fashion Assortment
Outfits for both Barbie and Ken: #2970 – #2971, #3619 – #3620
#2516 My First Barbie Doll outfit
#1503 My First Ken Doll outfit

Accessories: Hats.

Cardboard Insert: Press out and assemble a tote to carry doll and outfits.

YEAR	NUMBER	PRICE	TITLE	MAKER	VALUE
1994, 1991	**#2371**		**Barbie Paper Doll Deluxe Edition**	**Golden**	**Available**

Same as #1695-1 except for different cover, without pages to color, or the storage pocket. A tote is included.

Cover:
A photograph of Hawaiian Fun Barbie #5940 dressed in a red outfit.

Cardboard Insert:
Press out and assemble a tote to carry the doll and outfits.

#2371 *Barbie Paper Doll Deluxe Edition* (Hawaiian Fun).

YEAR	NUMBER	PRICE	TITLE	MAKER	VALUE
1994, 1992	**#2389**		**Barbie Paper Doll Deluxe Edition**	**Golden**	**Available**

The same as #1690-2 except for the different cover and without pages to color or the storage pocket. A tote is included.

Cover:

A photograph of Locket Surprise Barbie #2976 wearing the original doll outfit.

Cardboard Insert:

Press out and assemble a tote to carry the doll and outfits.

Back Cover:

A framed portrait of Locket Surprise Barbie to press out and assemble.

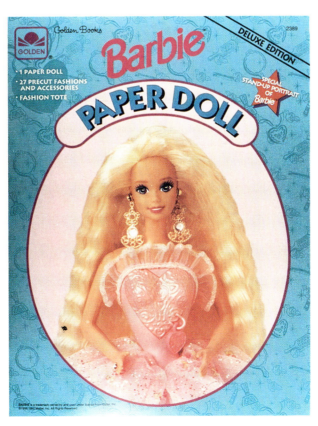

#2389 *Barbie Paper Doll Deluxe Edition* (Locket Surprise).

YEAR	NUMBER	PRICE	TITLE	MAKER	VALUE
1994, 1993	**#2748**		**Barbie Paper Doll Deluxe Edition**	**Golden**	**Available**

The same as #1502-4 except for two additional pages of outfits.

Cover:

J.C. Penney exclusive 1993 Happy Holiday Barbie stands at the foot of a staircase decorated for Christmas. Barbie is dressed in the original doll outfit.

Paper Doll:

One press-out doll: Fun to Dress Barbie #3240 in two-piece orchid lingerie.

Outfits:

Six pages of press-out outfits.

Accessories:

Hats and bows.

Cardboard Insert:

Press out and assemble a tote to carry the doll and outfits.

Back cover:

Press out and assemble a box to store trinkets or jewelry.

Peck-Gandré and Peck Aubry Paper Doll Books

The Peck Aubry Company (originally known as Peck-Gandré) has been producing extraordinary Barbie paper dolls since 1989. Although the Peck Aubry paper doll sets are a contemporary addition to Barbie doll's history, they feature outstanding replicas of vintage Barbie dolls, outfits, and accessories. In folios with outstanding color and accurate detail (often employing the photo real process), Peck Aubry has reproduced some of the most exceptional original Barbie dolls, including the Number One Barbie, the Bubble Cut, and the Swirl Ponytail.

Peck Aubry has also recreated Mattel's original wardrobes for these dolls, including the tiny buttons, satin linings, miniature zippers, sunglasses, open toe shoes, and dainty earrings. Peck Aubry folios include the fashions that reflect the careers which Barbie has pursued over the years, from airline stewardess to astronaut, registered nurse to doctor, fashion editor to business woman.

The first Peck Aubry Barbie fashions were introduced in 1989 to celebrate Barbie doll's 30th anniversary. This elegant collection of three folios included two "striped swimsuit" vintage Barbies and Ken and their wardrobes. Three additional folios were issued for Barbie doll's 35th anniversary; these featured a Number One, a Bubble Cut, and a Swirl Ponytail Barbie. Replicas of the American Girl, Color Magic, and Twist 'n Turn have recently been introduced.

"Our paper dolls are one of the few ways little girls can play with these exceptional dolls," commented Linda Peck, the owner and creative force behind Peck Aubry of Salt Lake City, Utah. "The original dolls and their clothes are now seen more often in doll collections than in the hands of children. With our Barbie folios, mothers and grandmothers can share their childhood with their daughters — Barbie paper dolls conjure up sweet memories for adults, while their children create new, exciting fantasy worlds."

YEAR	TITLE	MAKER	VALUE
1989	**Nostalgic Barbie Paper Doll**	**Peck-Gandré**	**$25.00**

The book is a two-sided, 9" x 12½" folio with one paper doll and six glossy sheets of outfits. The ready-to-dress doll is placed in the folio so her upper body can be seen through the oval window on the cover. The easy to handle outfit sheets include the original number and year of each outfit. The original outfits reproduced include the rare Gay Parisienne and Easter Parade which are beautiful and highly collectible. The attention to detail in the doll and the outfits is outstanding, from the color of Barbie doll's eyes to the many small accessories.

Cover:

A blonde #1 Ponytail Barbie paper doll shows through a clear plastic oval window on the cover. The border on the cover pictures ten small Barbie dolls in original outfits, #982 Solo in the Spotlight, #0876 Cheerleader, #964 Gay Parisienne, #967 Picnic Set, #939 Red Flare, #993 Sophisticated Lady, #984 American Airline Stewardess, #966 Plantation Belle, #971 Easter Parade and #962 Barbie-Q.

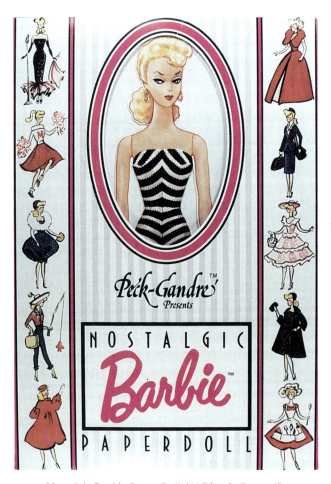

Nostalgic Barbie Paper Doll (#1 Blonde Ponytail).

Paper doll outfits #982 Solo in the Spotlight and #993 Sophisticated Lady from #1 Blonde Ponytail folio.

Paper Doll:

Blonde #1 Ponytail Barbie with pointed eyebrows, white irises, hoop earrings, and bright, red lips and nails wearing the original black and white striped swimsuit.

Outfits:

Three pages of cut-out outfits.

#0876 Cheerleader 1964	#971 Easter Parade 1959
#939 Red Flare 1962	#982 Solo in the Spotlight 1960
#942 Ice Breaker 1962	#984 American Airline Stewardess 1961
#964 Gay Parisienne 1959	#993 Sophisticated Lady 1963

To Color:

Three pages of cut-out outfits to color.

#941 Tennis Anyone 1962	#967 Picnic Set 1959
#944 Masquerade 1963	#972 Wedding Day 1959
#962 Barbie-Q 1959	#973 Sweet Dreams 1959
#966 Plantation Belle 1959	#976 Sweater Girl 1959

Accessories:

Tiara, hats, ice skates, megaphone, shoes with socks, tennis racket, tennis balls, slippers, clock, apple, diary, chef's hat, clown hat, and mask. Many other accessories are attached to the outfits.

Back cover:

"The Story of Barbie" relates a short history of Barbie and her family and gives information about the paper doll and outfits. A red heart sticker seals the folio.

Paper doll outfit Easter Parade #971 with 35th anniversary gift set outfit from #1 Blonde Ponytail folio.

YEAR	TITLE	MAKER	VALUE
1989	**Nostalgic Barbie Paper Doll**	**Peck-Gandré**	**$25.00**

The book is a two-sided, 9" x 12½" folio with one paper doll and six glossy sheets of outfits. The ready-to-dress doll is placed in the folio so her upper body can be seen through the oval window on the cover. The easy to handle outfit sheets include the original number and year of each outfit. The original outfits pictured, including the rare Roman Holiday, are beautiful and highly collectible. The attention to detail in the doll and outfits is outstanding, from the color of Barbie doll's eyes to the many accessories which complement the outfits beautifully and replicate the original outfits completely.

Cover:

A brunette #5 Ponytail Barbie paper doll can be seen through a clear plastic oval window in the cover. The borders on the cover picture ten small Barbies in original outfits, #993 Sophisticated Lady, #959 Theater Date, #968 Roman Holiday, #979 Friday Night Date, #934 After Five, #991 Registered Nurse, #0889 Candy Striper Volunteer, #961 Evening Splendor, #875 Drum Majorette, and #951 Senior Prom.

Paper Doll:

Brunette #5 Ponytail Barbie with curved eyebrows, blue irises, and pearl earrings wearing Fashion Queen original doll swimsuit #870.

Outfits:

Three pages of cut-out outfits.
#875 Drum Majorette 1964
#951 Senior Prom 1963
#961 Evening Splendor 1959
#968 Roman Holiday 1959
#979 Friday Night Date 1960
#983 Enchanted Evening 1960
#985 Open Road 1961

To Color:

Three pages of cut-out outfits to color.
#0889 Candy Striper 1964
#989 Ballerina 1961
#934 After Five 1962
#948 Ski Queen 1963
#959 Theater Date 1963
#969 Suburban Shopper 1959
#988 Singing in the Shower 1961
#987 Orange Blossom 1961
#991 Registered Nurse 1961

Nostalgic Barbie Paper Doll (#5 Brunette Ponytail).

Accessories:

Hats, bandanna, gloves, boots, hat with scarf and sunglasses, shoes, ballet shoes, tiara, boots, eyeglasses, diploma, hat, spoon, slippers, and shower hat.

Back Cover:

"The Story of Barbie" relates a short history of Barbie and her family and gives information about the paper doll and outfits. A red heart sticker seals the folio.

Paper doll outfits Enchanted Evening #983 and Evening Splendor #961 from #5 Brunette Ponytail folio.

#968 Roman Holiday outfit from 35th anniversary gift set with paper doll outfit from #5 Brunette Ponytail folio.

YEAR	TITLE	MAKER	VALUE
1989	**Nostalgic Ken Paper Doll**	**Peck-Gandré**	**$25.00**

The book is a two-sided, 9" x 12½" folio with one paper doll and six glossy sheets of outfits. The ready-to-dress doll is placed in the folio so his upper body can be seen through the oval window on the cover. The easy to handle outfit sheets include the original number and year of each outfit. The attention to detail in the doll and the outfits is outstanding, from the color of Ken doll's eyes to the many small accessories.

Cover:

A brunette #750 Ken paper doll can be seen through a clear plastic oval shape window on the cover. The border on the cover pictures five Barbie and Ken sets in original outfits, #979 Friday Night Date, #770 Campus Hero, #991 Registered Nurse, #793 Dr. Ken, #985 Open Road, #788 Rally Day, #944 Masquerade, #794 Masquerade, #875 Drum Majorette, #775 Drum Major, #876 Cheerleader, #799 Touchdown, #941 Tennis Anyone, #790 Time for Tennis, #984 American Airline Stewardess, #779 American Airline Captain, #942 Ice Breaker, #791 Fun On Ice, #951 Senior Prom and #787 Tuxedo.

Paper Doll:

#750 Ken dressed in his original red bathing trunks with yellow towel and sandals.

Outfits:

Three pages of cut-out outfits.
#0775 Drum Major 1964
#0779 American Airlines Captain 1964
#785 Dream Boat 1961
#788 Rally Day 1962
#790 Time for Tennis 1962
#791 Fun On Ice 1963
#798 Ski Champion 1963
#799 Touchdown 1963

To Color:

Three pages of cut-out outfits to color.
#7780 Campus Hero 1961
#784 Terry Togs 1961
#787 Tuxedo 1961
#793 Dr. Ken 1963
#794 Masquerade 1963
#796 Sailor 1963
#797 Army and Air Force 1963
#1319 Cheerful Chef 1964
#1409 Goin' Huntin' 1964

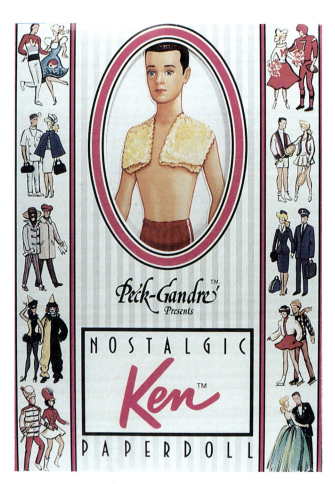

Nostalgic Ken Paper Doll.

Accessories:

Hats, helmet, knit caps, shoes, socks, baton, tennis shoes, clown mask, surgical mask, and cap with head mirror. Many other accessories are attached to the outfits.

Back cover:

"The Story of Barbie" relates a short history of Ken, his wardrobe, and his association with Barbie. A red heart sticker seals the folio.

YEAR	TITLE	MAKER	VALUE
1994	The 1959 (Number 1) Barbie Paper Doll	Peck Aubry	$20.00

The book is a two-sided, 9" x 12½" folio with one paper doll and two glossy sheets of outfits. The ready-to-dress paper doll is placed in the folio so her upper body can be seen through the oval window on the cover. The outfit tabs show the name, number, and year of the fashion. The folio reproduces many beautiful original outfits, including the rare Roman Holiday, Gay Parisienne and Easter Parade, which are highly collectible.

Cover:

A blonde #1 Ponytail Barbie paper doll can be seen through a clear plastic oval window on the cover. The pink border has pictures of many accessories, including sunglasses, #1 swimsuit, hat box, hangers, hats, gloves, purses, and compacts.

Paper Doll:

Photograph of blonde #1 Ponytail Barbie dressed in the original swimsuit.

Outfits:

One sheet divided into three sections with cut-out outfits.

#961 Evening Splendour 1959–1964 · #967 Picnic Set 1959–1961
#963 Resort Set 1959–1962 · #979 Friday Night Date 1960–1964
#964 Gay Parisienne 1959 · #982 Solo in the Spotlight 1960–1964
#966 Plantation Belle 1959–1961 · #983 Enchanted Evening 1960–1963

To Color:

One sheet divided into three sections with cut-out outfits to color.

#912 Cotton Casual 1959–1962 · #973 Sweet Dreams 1959–1963
#916 Commuter Set 1959–1960 · #975 Winter Holiday 1959–1963
#962 Barbie-Q 1959–1962 · #981 Busy Gal 1960–1961
#968 Roman Holiday 1959 · #986 Sheath Sensation 1961–1964
#971 Easter Parade 1959

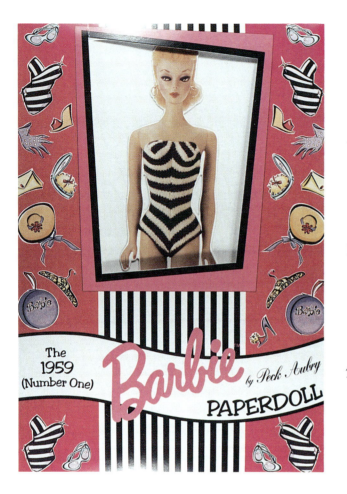

Accessories:

Hats, chef's hat, shoes, wedgies, slippers, pearls, charm bracelet, comb, hankie, eyeglasses and eyeglass case, picnic basket and fishing pole.

Back cover:

A short history of Barbie and information on the paper doll and outfits. A red heart sticker seals the folio.

The 1959 (Number One) Barbie Paper Doll.

YEAR	TITLE	MAKER	VALUE
1994	**The 1961 (Bubble Cut) Barbie Paper Doll**	**Peck Aubry**	**$20.00**

The book is a two-sided, 9" x 12½" folio with one paper doll and two glossy sheets of outfits. The ready-to-dress paper doll is placed in the folio so her upper body can be seen on the cover. The outfit tabs show the name, number, and year of the fashion.

Cover:

A brunette #850 Bubble Cut Barbie paper doll can be seen through a clear plastic oval window on the cover. A pink border has pictures of many accessories including sunglasses, #1 swimsuit, hat box, hangers, hats, gloves, purses, and compacts.

Paper Doll:

Photograph of brunette #850 Bubble Cut Barbie dressed in Fashion Queen original swimsuit #870.

Outfits:

One sheet divided into three sections with cut-out outfits.

#939 Red Flare 1962–1965
#954 Career Girl 1963–1964
#969 Suburban Shopper 1959
#984 Stewardess 1961–1964

#985 Open Road 1961–1962
#991 Registered Nurse 1961–1964
#993 Sophisticated Lady 1963–1964

To Color:

One sheet divided into three sections with cut-out outfits.

#850 Original Swimsuit 1962–1967
#931 Garden Party 1962–1963
#933 Movie Date 1962–1963
#934 After Five 1962–1964
#937 Sorority Meeting 1962–1963

#940 Mood For Music 1962–1963
#941 Tennis Anyone 1962–1964
#942 Ice Breaker 1962–1964
#0949 Stormy Weather 1964–1965

Accessories:

Hats, hat with attached scarf and sunglasses, tiara, shoes, wedgies, tennis shoes with socks, boots, halter, sunglasses, goggles, tennis rules book, and tennis balls. Many other accessories are attached to the outfits including #1613 Dogs 'n Duds 1964-1965.

Back Cover:

A short history of Barbie and information on the paper doll and outfits. A red heart sticker seals the folio.

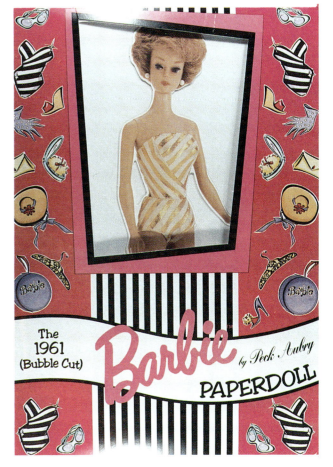

The 1961 (Bubble Cut) Barbie Paper Doll.

YEAR	TITLE	MAKER	VALUE
1994	The 1964 (Swirl Ponytail) Barbie Paper Doll	Peck Aubry	$20.00

The book is a two-sided, 9" x 12½" folio with one paper doll and two glossy sheets of outfits. The ready-to-dress paper doll is placed in the folio so her upper body can be seen through the oval window on the cover. The outfit tabs show the name, number, and year of the fashion.

Cover:

A dark hair #0850 Swirl Ponytail Barbie paper doll can be seen through a clear plastic oval window on the cover. A pink border has pictures of many accessories including sunglasses, #1 swimsuit, hat box, hangers, hats, gloves, purses, and compacts.

Paper Doll:

Photograph of a dark hair #0850 Swirl Ponytail Barbie dressed in the original swimsuit.

Outfits:

One sheet divided into three sections with cut-out outfits.
#0875 Drum Majorette 1964–1965
#0876 Cheerleader 1964–1965
#947 Bride's Dream 1963–1965
#948 Ski Queen 1963–1964
#1612 Theater Date 1964
#1641 Miss Astronaut 1965

To Color:

One sheet divided into three sections with cut-out outfits.
#0889 Candy Striper Volunteer 1964
#957 Knitting Pretty 1963
#1034 Hostess Set 1965
#1609 Black Magic Ensemble 1964–1965
#1615 Saturday Matinee 1965
#1622 Student Teacher 1965–1966
#1631 Aboard Ship 1965

The 1964 (Swirl Ponytail) Barbie Paper Doll

Accessories:

Hats, shoes, tennis shoes, eyeglasses, cake on tray, tray, knife, fork, spoon, scissors, bowl with yarn, globe, geography book, how to knit book, hot water bottle, tissue, nurse's cap, flower arrangement, two candleholders, toaster, tea kettle, coffee maker, three casseroles, place setting on placemat, teapot, and potholder.

Back Cover:

A short history of Barbie and information on the paper doll and outfits. A red heart sticker seals the folio.

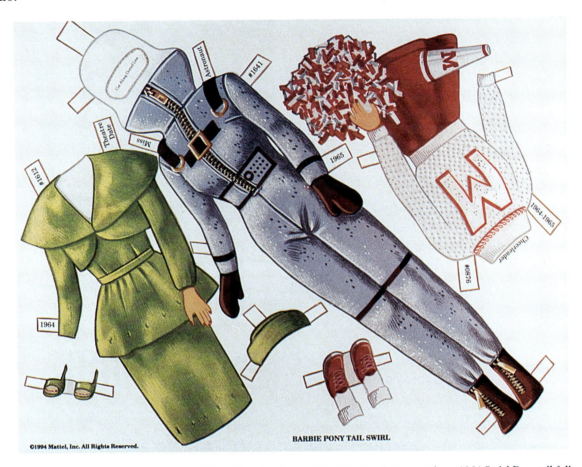

Paper doll outfits #1612 Theatre Date, #0876 Cheerleader, and #1641 Miss Astronaut from 1964 Swirl Ponytail folio.

YEAR	TITLE	MAKER	VALUE
1995	**The 1965 (American Girl) Barbie Paper Doll**	**Peck Aubry**	**Available**

The book is a two-sided, 9″ x 12½″ folio with one paper doll and two glossy sheets of outfits. The ready-to-dress doll is placed in the folio so her upper body can be seen through the oval window on the cover. The outfit tabs show the name, number, and year of the fashion. The excellent paper doll and beautiful outfits are highly collectible.

Cover:

A dark hair #1070 American Girl Barbie paper doll can be seen through a clear plastic oval window on the cover. A pink border has pictures of many accessories, sunglasses, #1 swimsuit, hat box, hangers, hats, gloves, purses, and compacts.

Paper Doll:

Photograph of a dark hair #1070 American Girl Barbie (bendable legs) dressed in a strapless terry cloth peach swimsuit.

Outfits:

One sheet divided into three sections with cut-out outfits.

#1625 Modern Art 1965
#1635 Fashion Editor 1965
#1645 Golden Glory 1965–1966
#1651 Beau Time 1966–1967
#1652 Pretty As a Picture 1966–1967
#1956 Fashion Luncheon 1966–1967
#1664 Shimmering Magic 1966–1967
#1668 Riding in the Park 1966–1967
#1678 Pan American Stewardess 1966

To Color:

One sheet divided into three sections with cut-out outfits.

#1640 Matinee Fashion 1965
#1642 Slumber Party 1965
#1643 Poodle Parade 1967
#1647 Golden Glamour 1965
#1661 London Tour 1966
#1633 Music Center Matinee 1966–1967
#1667 Benefit Performance 1966–1967
Sears Exclusive Pink Formal 1966

Accessories:

Hats, shoes, slippers, and bath scale.

Back Cover:

A short history of Barbie and information on the paper doll and outfits. A red heart sticker seals the folio.

The 1965 (American Girl) Barbie Paper Doll.

YEAR	TITLE	MAKER	VALUE
1996	**The 1966 (Color Magic) Barbie Paper Doll**	**Peck Aubry**	**Available**

The folio is a two-sided, (9" x 12½") folder holding one paper doll and two glossy sheets of outfits. The ready-to-dress paper doll is placed in the folio so her upper body can be seen through the oval window on the cover. The outfit tabs have the name, number, and year of the fashion. The accuracy in detail and color of all the outfits is outstanding. Magnificent collection of Sew-Free and Color Magic fashions.

Cover:

A blonde #1150 Color Magic Barbie paper doll can be seen through a clear plastic oval window on the cover. A pink border has pictures of many accessories including shoes, purses, gloves, scarves, hand mirror, hairbrush, and telephone.

Paper Doll:

Photograph of blonde #1150 Color Magic Barbie dressed in a one-piece diamond pattern swimsuit.

Outfits:

One sheet divided into three sections with cut-out outfits.

#1649 Lunch on the Terrace 1966–1967
#1650 Outdoor Art Show 1966–1967
#1666 Debutante Ball 1966–1967
#1676 Fabulous Fashion 1966

Sew-Free Fashions 1965–1966:
#1701 Sightseeing
#1707 Hootenanny
#1708 Patio Party
#1713 Sightseeing

To Color:

One sheet divided into three sections with cut-out outfits.
A Color Magic Costume 1966–1967:
#1775 Stripes Away
#1776 Smart Switch
#1777 Pretty Wild
#1778 Bloom Bursts
#1778 Bloom Bursts (different style)
#1779 Mix 'n Matchers (two outfits)
#4040 The Color Magic Fashion Designer Set dress

Accessories:

Hats, shoes, scarves, and attached purses.

Back cover:

A short history of Barbie and information on the paper doll and outfits. A red heart sticker seals the folio.

YEAR	TITLE	MAKER	VALUE
1996	**The 1967 (Twist 'n Turn) Barbie Paper Doll**	**Peck Aubry**	**Available**

The folio is a two-sided, (9" x 12½") folder holding one paper doll and two glossy sheets of outfits. The ready-to-dress doll is placed in the folio so her upper body can be seen through the oval window on the cover. The outfit tabs have the name, number, and year of the fashion. The excellent outfits, including the beautiful Floating Gardens gown with bracelets and earrings, beautifully reproduce the original outfits.

Cover:

A redhead #1160 Twist 'n Turn Barbie can be seen through a plastic oval window on the cover. The pink border pictures many accessories: shoes, purses, gloves, scarfs, hand mirror, hairbrush, and telephone.

Paper Doll:

Photograph of redhead #1160 Twist 'n Turn Barbie dressed in original two-piece swimsuit without the cover-up.

Outfits:

One sheet divided into three sections with cut-out outfits.

#1683 Sunflower 1967–1968   #1805 Bouncy Flouncy 1967–1968
#1690 Studio Tour 1967   #1809 Mini Print 1967–1968
#1695 Evening Enchantment 1967   #1810 Bermuda Holidays 1967–1968
#1696 Floating Gardens 1967   Sears Exclusive Weekenders 1967

To Color:

One sheet divided into three sections with cut-out outfits.

#1470 Intrigue 1967–1968
#1688 Travel Togethers 1967
#1460 Tropicana called #1683 Sunflower 1967–1968
#1686 Print Aplenty 1967–1968
#1687 Caribbean Cruise 1967
#1691 Fashion Shiner 1967–1968
#1806 Pajama Pow 1967–1968
#1807 Disco Dater 1967–1968

Accessories:

Shoes, hats, and earrings.

Back cover:

A short history of Barbie and information on the paper doll and outfits. A red heart sticker seals the folio.

Barbie Doll's Friends in Paper Doll Books

Many of Barbie doll's friends have appeared in their own paper doll books, including celebrities such as the London model Twiggy, popular in the 1960s , and television personalities Buffy and her doll, Mrs. Beasley (from the series, *Family Affair*) and Donny and Marie Osmond. Mrs. Beasley had several of her own books while Donny and Marie had their original doll wardrobes duplicated in their paper doll book. The characters from the movie *Chitty Chitty Bang Bang*, including Truly Scrumptious, also had their own book. Other paper doll books featured Barbie doll's friends Miss America, the Heart Family, and Barbie doll's high school friends, Starr, Shaun, Kelley, and Tracy.

Back cover #1526 *Heart Family Paper Doll*

YEAR	NUMBER	PRICE	TITLE	MAKER	VALUE
1967	**#1999**	**$1.00**	**Twiggy Paper Doll**	**Whitman**	**$45.00**

A great collection of the '60s mod-fashions with Barbie's celebrity friend, Twiggy, the well-known London model. A look back to the '60s.

Cover:
A large close-up of Twiggy on a shocking pink background with colored circles. Three small Twiggys hold paper dolls.

Paper Doll:
One press-out doll: Twiggy #1185 wears a strapless shocking pink swimsuit with an orange laced see-through insertion at each side of the waist.

Outfits:
Six pages of press-out outfits.

Inside Front Cover:
Seven circular shapes hold Twiggy photographs.

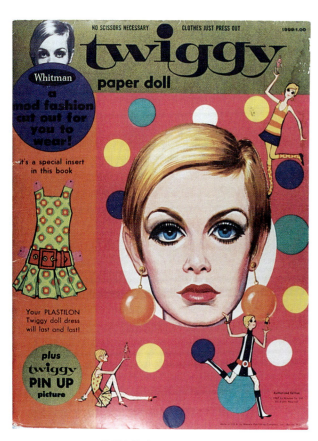

#1999 *Twiggy Paper Doll.*

Twiggy paper doll from #1999 book.

Plastilon Insert:
A cut-out child-sized plastilon mod dress like Twiggy's to wear.

Cardboard Insert:
A pin-up photograph of Twiggy.

Inside Back Cover:
Various photographs of Twiggy in five circles and two rectangles.

Back Cover:
A black and white photograph of Twiggy against a shocking pink background with blue, green, yellow, and white circles.

YEAR	NUMBER	PRICE	TITLE	MAKER	VALUE
1968	#1995	69¢	**Buffy Paper Doll**	Whitman	$40.00

Buffy is one of Barbie doll's celebrity friends. A charming paper doll book with a good size paper doll that is fun to dress.

#1995 *Buffy Paper Doll* (with Mrs. Beasley).

Cover:

Photograph of Anissa Jones who played "Buffy" on the CBS *Family Affair* television program holding her Mrs. Beasley doll. The cover also features drawings of Buffy #3577 and some of her clothes and a pink duck.

Paper Doll:

Two press-out dolls: Large photograph of Buffy #3577 dressed in black leotards and a tiny Mrs. Beasley #3577 doll.

Outfits:

The outfits are on press-out pages. Additional information is not currently available because an uncut book has not been located.

Inside Cover:

Large Buffy standing beside her dresser which has a handy carry-pocket to store dolls and outfits

Back Cover:

Three beautiful photographs: Buffy talking on the phone and holding Mrs. Beasley doll, Buffy showing her dress while dragging Mrs. Beasley doll, and a close-up of Mrs. Beasley.

YEAR	NUMBER	PRICE	TITLE	MAKER	VALUE
1968	#1982	59¢	Chitty Chitty Bang Bang	Whitman	$40.00

Barbie as Truly Scrumptious was released in 1969. The paper doll book appeared before the Barbie doll.

Cover:

Truly Scrumptious, Potts, and the flying Chitty Chitty Bang Bang car with the children.

Paper Dolls:

Five press-out paper dolls. Truly is dressed in a one-piece pink swimsuit trimmed with lace and ribbon. Potts wears a two-piece white swimsuit with wide horizontal red stripes. Grandpa wears one-piece short underwear, red socks with garters, and high top shoes. Jeremy wears blue and white shorts and white T-shirt with high top button shoes. Jemima wears a short slip in yellow with lace trim.

Outfits:

Six press-out pages of outfits. The outfits are replicas of the clothes worn in the movie, *Chitty Chitty Bang Bang.*

Inside Front Cover:

Four drawings of scenes from the movie inside clouds floating over houses. The cover has a carry-pocket to store dolls.

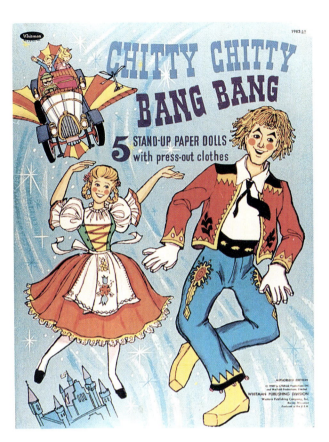

Back Inside Cover:

Four drawings of scenes from the movie inside clouds floating over houses. The cover has a carry-pocket to store clothes.

Back Cover:

Truly Scrumptious, Potts, Jemima, and Jeremy in the car flying over the rooftops.

#1982 Chitty Chitty Bang Bang

YEAR	NUMBER	PRICE	TITLE	MAKER	VALUE
1969	#1985	59¢	**Buffy Paper Doll**	Whitman	$35.00

Buffy is one of Barbie's celebrity friends. The book includes three Buffy photographs on the outfits pages to fold and place in a wallet.

Cover:

Photographs of Anissa Jones, who played "Buffy" on the CBS television program. A large close-up photograph of Buffy is in the center surrounded by photographs of Buffy #3577 dressed in a fur coat and hat, holding a muff, and Buffy modeling a blue dress.

Paper Dolls:

Two press-out dolls: Buffy #3577 dressed in black leotards and Mrs. Beasley #3577.

Outfits:

Six pages of press-out outfits. Thirteen Buffy outfits for every occasion, five outfits for Mrs. Beasley.

Accessories:

Hats, scarf, necklace, and bag.

#1985 *Buffy Paper Doll.*

Buffy paper doll from #1985 book.

Inside Cover:
  Press out and assemble as tote for carrying dolls and outfits.

Cardboard Insert:
  Mrs. Beasley has a small tote to press out.

Back Cover:
  A beautiful photograph of Buffy modeling a blue dress with white lace tights and shoes and holding Mrs. Beasley at her side.

YEAR	NUMBER	PRICE	TITLE	MAKER	VALUE
1970	#1973	39¢	Mrs. Beasley Paper Doll Book	Whitman	$25.00

A charming paper doll book featuring Mrs. Beasley, who was the rag doll of Buffy Davis on the CBS television program Family Affair.

Cover:

Mrs. Beasley #3577 doll standing in the garden holding a sign which reads, "Mrs. Beasley Paper Doll Book."

Paper Doll:

One press-out doll: full-cover drawing of Mrs. Beasley #3577, as large as the book, dressed in the original blue swimsuit with white dots.

Outfits:

Six pages of press-out outfits. Thirteen outfits for all occasions.

Accessories: None.

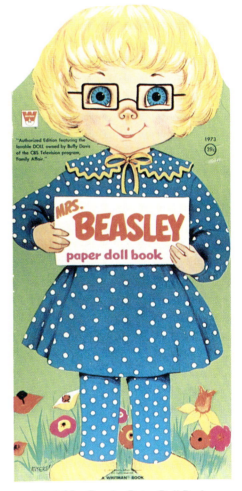

#1973 *Mrs. Beasley Paper Doll Book.*

YEAR	NUMBER	PRICE	TITLE	MAKER	VALUE
1972	#1986	69¢	Mrs. Beasley Paper Doll Fashions	Whitman	$20.00

One of Barbie doll's celebrity friends in a very special book. The wardrobe and dressing table add a new segment to this paper doll book.

Cover:

Mrs. Beasley #3577 sitting in a rocking chair.

Paper Doll:

One press-out doll: Mrs. Beasley #3577 wears flowered undies.

Outfits:

Six pages of press-out outfits. Mrs. Beasley #3577 original outfit and other outfits for all occasions.

Accessories: Hats.

#1986 *Mrs. Beasley Paper Doll Fashions.*

Mrs. Beasley paper doll from #1986.

Cardboard Insert:
Large bed for Buffy with slits at the edge for inserting a press-out cloth dog.

Inside Front Cover:
A pink dressing table with a carry-pocket for storing doll and outfits.

Inside Back Cover:
A pink wardrobe with slits for inserting a press-out clock and framed photo of Buffy. The cover has a carry-pocket for storing doll and outfits.

Back Cover:
Five pictures of Mrs. Beasley modeling outfits from the book.

YEAR	NUMBER	PRICE	TITLE	MAKER	VALUE
1972	#1993	79¢	**Mrs. Beasley Paper Doll Fashions**	Whitman	$20.00

The same as #1986 except for the different stock number and price.

YEAR	NUMBER	PRICE	TITLE	MAKER	VALUE
1973	#1978	69¢	**Miss America Paper Doll**	Whitman	$20.00

A good collectible for Miss America fans. The wardrobe and paper doll are very well done.

Cover:
A photograph of Miss America 1972, Laura Lee Schaefer, holding a bouquet and wearing a beautiful white gown, banner, and crown.

Paper Doll:
One press-out doll: The brunette Miss America #3194-9991 (Kellogg Company offer) wears a one-piece, light blue swimsuit.

Outfits:
Six pages of press-out outfits. A variety of beautiful outfits for Miss America.

Inside Front Cover:
Two contestants stand on a stage beside Miss America's throne.

Cardboard insert:

A runway to press out and assemble.

Inside Back Cover:

Three contestants wearing beautiful gowns with banners stand on the stage. The cover has a carry-pocket for storing clothes.

Back Cover:

A profile photograph of Miss America, Laura Lee Schaefer.

#1978 *Miss America Paper Doll.*

YEAR	NUMBER	PRICE	TITLE	MAKER	VALUE
1977	#1991	79¢	**Donny and Marie Paper Dolls**	Whitman	$20.00
			A nice collection of beautiful matching Donny and Marie outfits.		

Cover:

Photographs of Donny and Marie appear in stars on a blue background.

Paper Dolls:

Two press-out paper dolls: Marie #9768 is dressed in a one-piece yellow and white swimsuit and #9767 Donny wears a matching yellow shirt and white trunks.

Outfits:

Four pages of press-out outfits.

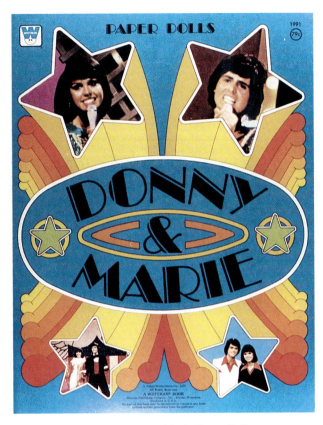

#1991 *Donny and Marie Paper Dolls.*

Donny and Marie paper dolls from #1991 book.

Outfits (continued):

Marie:
#2450 Glimmer O' Gold
#2451 Silver 'n Shine
#2452 Country Hoedown
#2492 Peasant Sensation

Donny:
#9813 Deepest Purple
#9814 South O' the Border (different colors)
#9815 Silver Shimmer
#9816 Starlight Night

Cardboard Insert:

Press out and assemble stage for Donny and Marie.

Back Cover:

Large oval photograph of Donny and Marie singing into microphones. Their names appear in two large stars above them.

YEAR	NUMBER	PRICE	TITLE	MAKER	VALUE
1979	#1976-3	89¢	**Miss America Paper Doll**	Whitman	$20.00

A good collectible for Miss America fans. The wardrobe and paper doll are very well done.

Cover:

A full-length photograph of Miss America 1972, Laura Lee Schaefer.

Paper Doll:

One press-out doll: smiling brunette Miss America dressed in blue one-piece swimsuit tied at the waist.

Outfits:

Four pages of press-out outfits.

Cardboard Insert:

Press out and assemble a red, white, and blue runway.

Back Cover:

A profile photograph of Miss America, Laura Lee Schaefer.

#1976-3 *Miss America Paper Doll.*

YEAR	NUMBER	PRICE	TITLE	MAKER	VALUE
1980	#1982-31	99¢	**Starr Paper Dolls**	Whitman	$20.00

Starr and her high school friends are considered by some collectors to be Barbie doll's classmates.

Cover:

Starr wears her original doll outfit with short sleeves while talking on the phone. To the side of Starr on the green background are large pictures of Tracy, Kelley, Shaun, and Starr.

Paper Dolls:

Four press-out dolls: Starr #1280 wears pink shorts and top. Shaun #1283 wears a red T-shirt with a large "S" on the front with red trunks. Kelley #1281 wears green shorts and top. Tracy #1282 wears blue shorts and top.

Outfits:

Four pages of press-out outfits.

#1280 Starr original doll outfit   #1282 Tracy original doll outfit
#1281 Kelley original doll outfit   #1283 Shaun original doll outfit

#1982-31 *Starr Paper Dolls.*

Starr doll and Starr paper doll from #1982-31 book.

Outfits (continued):

Shaun, Tracy, Kelley
 #1388 Casual Living
 #1389 Cheerleader Captain
 #1393 Homecoming Queen
 #1394 Disco Dancing

Shaun
#1395 Prom Night
#1398 School Days

Accessories: None.

Back Cover:
 Each doll from the Springfield High School group (Kelley, Starr, Tracy, and Shaun) holds an instrument.

YEAR	NUMBER	PRICE	TITLE	MAKER	VALUE
1980	#1982-41	$1.29	Starr Paper Dolls	Whitman	$20.00

The same as #1982-31 except for the price.

YEAR	NUMBER	PRICE	TITLE	MAKER	VALUE
1981	#1836-42	$2.59	Starr A Paper Doll Playbook	Golden	$20.00

The playbook has sundaes, sodas, lost and found items, and signs to place in the scene on the inside cover. Starr and her friends share Barbie doll's beauty and style and are generally accepted by many to be her close friends.

Cover:
 Starr and Shaun wear their original doll outfits and share an ice cream soda with two straws. Three small square pictures at the bottom of the cover show items in the book, including the dolls and clothes, the dance play scene, and the classroom play scene.

Paper Dolls:
 Four press-out dolls: Starr #1280 wears a white top and shorts decorated with a star. Shaun #1283 wears blue trunks and the original doll top. Tracy #1282 wears blue shorts and a yellow top with red and white stripes. Kelley #1281 wears green shorts and a white top trimmed with a green bow.

Outfits:
 Four pages of press-out outfits.

Outfits (continued):

#1280 Starr original doll outfit
#1281 Kelley original doll outfit
#1282 Tracy original doll outfit (two)
#1283 Shaun original doll outfit
#1392 Prom Night

#1393 Homecoming Queen
#1394 Disco Dancing
#1388 Casual Living
#1389 Cheerleader Captain

Shaun
#1395 Prom Night
#1398 School Days

Accessories:

Crown for Homecoming Queen.

Cardboard Insert:

Soda fountain table and two chairs, stage, instruments, and books.

Inside Front Cover:

A group of five Springfield High group dance at the soda fountain.

Inside Back Cover:

Unfold the cover to reveal Shaun dancing at the soda fountain.

Back cover:

The unfolded cover reveals the teacher and students at school with lockers and a bulletin board.

#1836-42 *Starr A Paper Doll Playbook.*

YEAR	NUMBER	PRICE	TITLE	MAKER	VALUE
1985	#1526	$1.29	Heart Family Paper Doll	Golden	$20.00

A charming book with smiling faces and tiny pink hearts decorating the paper doll pages.

Cover:

The #9439 Heart Family Deluxe Set, Mom with Baby Girl and Dad with Baby Boy, wearing the original doll outfits.

Paper Dolls:

Four press-out dolls: Heart Family Deluxe Set #9439 dolls with Mom wearing a one-piece undergarment, Baby Girl in a top and shorts carrying a doll, Dad wearing a white T-shirt and blue trunks, and Baby Boy in shorts and top holding a ball.

Outfits:

#9439 Mom, Dad, and Babies original doll outfits.
Mom and Baby Girl Fashions: #9593, #9594, #9595, #9596
Dad and Baby Boy Fashions: #9597

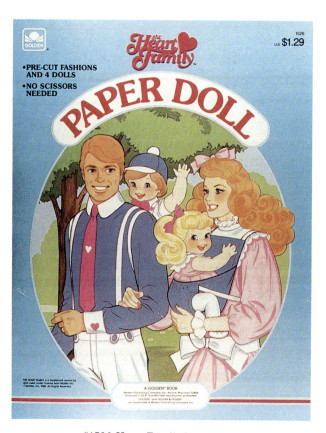

#1526 *Heart Family Paper Doll.*

Heart Family Mom and Baby fashion "Nightgown" with matching paper doll nightgowns from #1526 book.

Accessories:
   One little hat.

Back Cover:
   Dad on his knees with Baby Girl and Baby Boy on his back while Mom is trying to take their picture with a camera.

YEAR	NUMBER	PRICE	TITLE	MAKER	VALUE
1985	**#1526**		**Heart Family Paper Doll**	**Golden**	**$20.00**
			The same as #1526 except no price.		

Barbie and Family Paper Doll Books and Folders
Whitman - Golden

Year	Title	Code Number
1962	Barbie and Ken Cut-Outs (folder)	1971
	Barbie and Ken Cut-Outs	1963
	Barbie Doll Cut-Outs (folder)	1963
	Barbie Doll Cut-Outs (reprint #1963)	1957
	Barbie Doll Cut-Outs (reprint #1963)	1966
	Barbie Doll Cut-Outs (folder)	1962
1963	Barbie Cut-Outs (folder)	1962
	Midge, Barbie's Best Friend Cut-Outs	1962
	Barbie and Ken Cut-Outs (folder)	1976
	Barbie, Ken and Midge Paper Dolls (folder)	1976
1964	Barbie Costume Dolls with Skipper, Ken, Midge and Allan (folder)	1976
	Barbie and Skipper (folder)	1957
	Barbie and Skipper (folder, reprint #1957)	1944
	Barbie and Skipper (reprint #1957)	1944
1965	Skipper – Barbie's Little Sister	1984
	Skooter Paper Dolls (folder)	1985
1966	Barbie, Skipper and Skooter Paper Dolls (folder)	1976
	Meet Francie, Barbie's 'Mod'ern Cousin Paper Dolls (folder)	1980
1967	Barbie Has A New Look! Paper Dolls (folder)	1976
	Barbie Has A New Look! Paper Dolls (reprint #1976)	1996
	Francie, Barbie's 'Mod'ern Cousin & Casey, Francie's New Friend (folder)	1986
1968	Barbie, Christie, Stacey Paper Dolls (folder)	1976
	Barbie, Christie, Stacey Paper Dolls (folder, reprint #1976)	1978
	Tutti Paper Doll, Barbie and Skipper's Tiny Sister	1991
1969	Barbie Dolls and Clothes (folder)	1976
1970	Barbie and Ken Paper Dolls	1976
	Barbie and Ken Paper Dolls (reprint #1976)	1985
	Barbie and Ken Paper Dolls (reprint #1976)	1986
	New 'n Groovy P.J. Paper Doll	1981
1971	Groovy World of Barbie and Her Friends Paper Dolls (folder)	1976
	P.J. Cover Girl Paper Doll	1981
	World of Barbie Paper Dolls (folder)	1987
1972	Groovy P.J. Paper Doll Fashions	1974
	Pos'n Barbie Paper Doll Fashions	1975
	Malibu Barbie, The Sun Set (six pages)	1994
	Malibu Barbie, The Sun Set (four pages)	1994

Year	Title	Code Number
1973	Barbie's Boutique (folder)	1954
	Barbie's Boutique (reprint with price change)	1954
	Barbie's Boutique	1947
	Barbie's Boutique (reprint with price change)	1947-1
	Barbie's Boutique	1996-1
	Quick Curl Barbie And Her Paper Doll Friends (folder, six pages)	1984
	Quick Curl Barbie And Her Paper Doll Friends (reprint, four pages)	1984
	Barbie Country Camper Doll Book	1983
	Barbie Country Camper Doll Book (folder)	1990
	Barbie's FriendShip Paper Dolls	1996
	Malibu Francie Doll Book (folder)	1955
	Malibu Francie Doll Book	1955
	Francie With Growin' Pretty Hair (folder and book)	1982
	Hi! I'm Skipper Paper Doll Fashions	1969
	Malibu Skipper	1945-2
	Malibu Skipper (folder)	1952
	Malibu Skipper (reprint with price change)	1952
	Malibu Skipper (folder, reprint with encircled price)	1952
	Malibu Skipper (reprint)	1952
1974	Barbie Goin' Camping	1951
	Barbie Goin' Camping (reprint with price change)	1951
	Barbie Goin' Camping	1961
	Barbie's Sweet 16 Paper Dolls	1981
1975	Yellowstone Kelley Paper Dolls	1956
	Barbie and Her Friends All Sports Tournament	1981
1976	Barbie Fashion Originals Paper Doll	1989
	Growing Up Skipper Paper Dolls	1990
	Barbie's Beach Bus Paper Dolls	1996
	Barbie's Beach Bus Paper Dolls (reprint with price change)	1996-1
1977	SuperStar Barbie Paper Doll	1983
	SuperStar Barbie Paper Doll (reprint with price change)	1983-2
	Ballerina Barbie Paper Dolls	1993
	Ballerina Barbie Paper Dolls	1993-1
1978	Fashion Photo Barbie And P.J.	1982-32
	Fashion Photo Barbie And P.J. (reprint with price change)	1982-32
	Fashion Photo Barbie And P.J.	1997
	Fashion Photo Barbie And P.J.	1997-1
	Fashion Photo Barbie And P.J. (reprint)	1997-21

Year	Title	Code Number
1979	Fashion Photo Barbie And P.J.	1997-21
1980	Barbie and Skipper Campsite at Lucky Lake Playbook	1836
	Barbie and Skipper Campsite at Lucky Lake Playbook (reprint #1836)	1836-31
	Barbie and Skipper Campsite at Lucky Lake Playbook (reprint #1836)	1836-41
	Barbie and Skipper Campsite Playbook (reprint #1836 with price change)	1836-41
	Super-Teen Skipper Paper Doll	1980-1
	Super-Teen Skipper Paper Doll	1980-3
	Super-Teen Skipper Paper Doll (reprint #1980-3)	1982-33
1981	Pretty Changes Barbie Paper Doll	1982-34
	Pretty Changes Barbie (reprint #1982-34)	1982-42
	Pretty Changes Barbie (reprint #1982-34 with cover and price change)	1982-42
1982	Western Barbie Paper Doll	1982-43
	Golden Dream Barbie Paper Doll	1983-43
1983	Sunsational Malibu Barbie Paper Doll	1982-44
	Angel Face Barbie Paper Doll	1982-45
	Twirly Curls Barbie Paper Doll	1982-46
	Pink and Pretty Barbie Paper Doll Playbook	1836-43
	Pink and Pretty Barbie Paper Doll	1983-44
1984	Barbie and Ken Paper Doll	1527
	Barbie and Ken Paper Doll (reprint #1527)	1985-51
	Barbie Christmas Time Paper Doll	1731
	Barbie Fantasy Paper Doll	1982-47
	Crystal Barbie Paper Doll	1983-46
	Great Shape Barbie Paper Doll	1982-49
	Great Shape Barbie Paper Doll (reprint #1982-49)	1985-51
1985	Great Shape Barbie Paper Doll	1522
	Great Shape Barbie Paper Doll	1525
	Great Shape Barbie Paper Doll	1982-49
	Great Shape Barbie Paper Doll	1985-51
	Peaches 'n Cream Barbie Paper Doll	1525
	Peaches 'n Cream Barbie Paper Doll	1983-48
	Day-to-Night Barbie Paper Doll	1521
	Day-to-Night Barbie Paper Doll (reprint #1521)	1982-48
1986	Barbie And The Rockers Paper Dolls	1528
	Barbie And The Rockers Paper Dolls (reprint #1528)	1528-1
	Tropical Barbie Paper Dolls	1523
	Tropical Barbie Paper Dolls (reprint #1523)	1523-1
1987	Jewel Secrets Barbie Paper Dolls	1537
	Jewel Secrets Barbie Paper Dolls (reprint #1537)	1537-1
1988	Perfume Pretty Paper Dolls	1500
1989	SuperStar Barbie Paper Doll	1537-2
1990	Barbie Paper Doll	1502

Year	Title	Code Number
1990	Barbie Paper Doll	1502-1
	Barbie Paper Doll	1523-2
	Barbie Paper Doll	1502-5
	Deluxe Paper Doll Barbie	1690
	Deluxe Paper Doll Barbie	1690-1
1991-90	Barbie Paper Doll (reprint #1502)	1502-2
1991	Deluxe Paper Doll Barbie	1695
	Deluxe Paper Doll Barbie	1695-1
1991-89	Barbie Paper Doll (reprint #1537-2)	1537-3
1991-90	Deluxe Paper Doll Barbie (reprint #1690)	1690-1
1992	Barbie Paper Doll	1502-3
	Deluxe Paper Doll Barbie	1690-2
1993	Barbie Paper Doll	1502-4
1993-90	Barbie Paper Doll (reprint #1502-1)	1502-5
1994	Barbie Paper Doll Deluxe Edition	2018
1994-92	Barbie Paper Doll Deluxe Edition (reprint #1695-1)	2371
	Barbie Paper Doll Deluxe Edition (reprint #1690-2)	2389
1994-93	Barbie Paper Doll Deluxe Edition	2748

Peck-Gandré and Peck Aubry Barbie Paper Dolls

Year	Title
1989	Nostalgic Barbie Paper Doll (Blonde) # 1 Ponytail
	Nostalgic Barbie Paper Doll (Brunette) #5 Brunette
	Nostalgic Ken Paper Doll
1994	The 1959 (Number One) Barbie Paper Doll
	The 1961 (Bubble Cut) Barbie Paper Doll
	The 1964 (Swirl Ponytail) Barbie Paper Doll
1995	The 1965 (American Girl) Barbie Paper Doll
1996	The 1966 (Color Magic) Barbie Paper Doll
	The 1967 (Twist 'n Turn) Barbie Paper Doll

Barbie's Friends Whitman and Golden Paper Doll Books

Year	Title	Code Number
1967	Twiggy Paper Doll	1999
1968	Buffy Paper Doll	1995
	Chitty Chitty Bang Bang	1982
1969	Buffy Paper Doll	1985
1970	Mrs. Beasley Paper Doll Book	1973
1972	Mrs. Beasley Paper Doll Fashions	1986
	Mrs. Beasley Paper Doll Fashions	1993
1973	Miss America Paper Doll	1978
1977	Donny and Marie Paper Doll	1991
1979	Miss America Paper Doll	1976-3
1980	Starr Paper Dolls	1982-31
	Starr Paper Dolls (reprint #1982-31)	1982-41
1981	Starr A Paper Doll Playbook	1836-42
1985	Heart Family Paper Dolls	1526
	Heart Family Paper Dolls	1526-1
	Heart Family Paper Dolls	1983-50

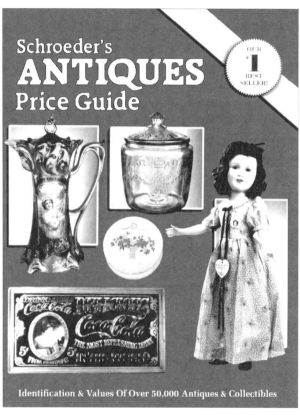